W9-BCT-455

# Praise for *Barn Club*

'A joyful reminder of why nature, being outside, being together and creating beauty is so good for the soul.'

—**Kate Humble**, broadcaster and author of *A Year of Living Simply*

'For the reader who wishes to resist the gathering pace of modern life and take time to learn from the past, the tale of hand-raising a barn the old-fashioned way brings nature, community and craftsmanship together in an enduring and satisfying feeling of a job well done.'

—**Gillian Burke**, co-presenter of BBC's *Springwatch*, writer and biologist

'In today's ego-techno-centred world, Robert Somerville's tale of elm trees, hand tools, timber framing and comradery is a welcome relief. His "Barn Club" approach is a way forward that utilises local traditions, local materials and local hands to create a built environment that is more harmonious with the natural world and of course more beautiful. Now, if every community around the world had one of these Barn Clubs, how nice would that be?'

—**Jack A. Sobon**, architect, timber framer and author of *Hand Hewn*

'Elm trees may have been devastated by Dutch elm disease but they are still with us and should not be forgotten, as Robert Somerville powerfully shows. Natural history, ancient crafts and a group of twenty-first-century volunteers meet in this book to show us how elms can reconnect us to nature, past cultures and one another. A beautiful and timely book with a barnful of good ideas.'

—**Professor Richard Buggs**,
Royal Botanic Gardens, Kew

'Robert Somerville is to be congratulated for his understanding of the entomology and pathology involved in Dutch elm disease, which in addition to its devastation of British elms also resulted in the loss of over 300 million American elms. His book should find a ready readership among do-it-yourselfers, whose home improvement projects have multiplied in this Covid environment.'

—**John Hansel**, founder,
Elm Research Institute

# Barn Club

## A Tale of Forgotten Elm Trees, Traditional Craft and Community Spirit

Robert J. Somerville

Chelsea Green Publishing
White River Junction, Vermont
London, UK

Commissioning Editor: Jonathan Rae
Project Editor: Benjamin Watson
Copy Editor: Susan Pegg
Proofreader: Nikki Sinclair
Designer: Melissa Jacobson
Page Layout: Abrah Griggs

Printed in the United States of America.
First printing February 2021.
10 9 8 7 6 5 4 3 2 1 21 22 23 24 25

**Our Commitment to Green Publishing**

Chelsea Green sees publishing as a tool for cultural change and ecological stewardship. We strive to align our book manufacturing practices with our editorial mission and to reduce the impact of our business enterprise in the environment. We print our books and catalogs on chlorine-free recycled paper, using vegetable-based inks whenever possible. This book may cost slightly more because it was printed on paper that contains recycled fiber, and we hope you'll agree that it's worth it. *Barn Club* was printed on paper supplied by Sheridan that is made of recycled materials and other controlled sources.

**Library of Congress Cataloging-in-Publication Data**

Names: Somerville, Robert J., author.
Title: Barn Club : a tale of forgotten elm trees, traditional craft and community spirit / Robert J. Somerville.
Description: White River Junction, Vermont : Chelsea Green Publishing, [2021] | Includes bibliographical references.
Identifiers: LCCN 2020049415 (print) | LCCN 2020049416 (ebook) | ISBN 9781603589666 (hardback) | ISBN 9781603589673 (ebook)
Subjects: LCSH: Barns--England--Hertfordshire--Design and construction. |Vernacular architecture--England--Hertfordshire. | Building, Wooden--England--Hertfordshire. | Elm.
Classification: LCC TH4930 .S66 2021 (print) | LCC TH4930 (ebook) | DDC 725/.372094258--dc23
LC record available at https://lccn.loc.gov/2020049415
LC ebook record available at https://lccn.loc.gov/2020049416

Chelsea Green Publishing
85 North Main Street, Suite 120
White River Junction, Vermont USA

Somerset House
London, UK

www.chelseagreen.com

*To unknown village carpenters*
*and elm trees, everywhere.*

# Contents

The Great Barn at Wallington, viewed from the lane.
*With kind permission of Liz Somerville.*

# Preface

The discovery of an ancient barn made of elm wood in Wallington near where I live set off a series of questions about what being a carpenter was like before the Industrial Revolution. The big idea was to follow in the footsteps of these unnamed craftsmen of yore, to try and recreate the experience of making a barn by hand.

My clients, Bruce and Esther Carley, were volunteers on a previous barn project and commissioned one for themselves at Churchfield Farm, their own smallholding in the Hertfordshire village of Tewin. From the beginning, volunteers were involved. This became known as Barn Club.

The Carley Barn was made and raised by hand, inspired by the eighteenth-century carpentry traditions of Hertfordshire. In spite of Dutch elm disease, all the elm for the post-and-beam timber frame was sourced from local woods. This book closely follows the story of the trees; from the woodland to the yard where the logs were milled, where the joints were scribed and cut with ancient hand tools, and where eventually the barn was raised by hand with the help of forty-eight people.

The village carpenters of generations ago would have had a high degree of skill with hand tools and an incredible depth of knowledge of their local trees that is now almost completely forgotten. Some of that knowledge is laid out in the chapters of this book.

The following pages are written to enlighten and inspire the reader about elm trees, traditional carpentry using hand tools, and the joys of working together as a team, outside in the fresh air.

# Acknowledgements

This book would not have been possible were it not for my wife Lydia, whose encouragement and reflections have steered it throughout. Equal thanks go to Jon Rae and Margo Baldwin of Chelsea Green Publishing, for spotting a book within the story of building a barn, to Ben Watson, Matt Haslum, Susan Pegg and Muna Reyal for developing and shaping the manuscript, and to Rosie Baldwin, Katie Read, Pati Stone and Neil Gower for preparing *Barn Club* for publication.

The barn we built in Tewin, Hertfordshire, is only there at all thanks to Esther and Bruce Carley and their vision to commission something really special in the landscape they love. Thanks, of course, to all of the Barn Club volunteers and barn raisers, young and old, for their goodwill, humour and general grit to make it all happen. A big thanks to Joel Hendry for his advice, experience and companionship throughout the project, and also to the professional timber framers who came to help out at crucial moments.

Thank you to my sister, Liz, for her drawing and my oldest daughter, Phoebe, also helped with reading through manuscripts and looking over drawings, noting when things didn't make sense. Many of the photographic images were taken by Jonathan Sampson and a promotional video was filmed by Steve Robson. They both kindly allowed their work to be reproduced. Thanks should also be given to the

Woodhall Estate at Watton-at-Stone, who provided the elm trees, and the foresters in the distant past who planted them.

Casting a wider net, I would like to acknowledge all the people who have influenced me through the years of my learning to make things by hand, notably Joe Thompson, Joe Hawkswell and the innumerable people who shared their skills and showed me how to use a particular tool or master a technique. These include my younger brother Matt and, when first starting out, my business partner, Richard Vickerman. I should also note that patrons create opportunity for craftspeople, so I would not have begun at an early age had it not been for my stepfather John Henderson who commissioned a barn from me in 1978. Beyond these people are cohorts of craftsmen and craftswomen over the centuries, who knew their materials well and loved their work. To these custodians of a way of living, time beyond mind, who have honed their skills and passed them on, thank you.

ONE

# A Different Way of Working

Craft is an event as much as a product. It is something that happens. What remains is both in the beauty of the object and in the hearts of those who took part. Barn Club is the epitome of this sentiment; a group of crafts-people and volunteers coming together to create something bigger than the individuals who made it, that will be standing long after they have gone but will keep a part of their story alive. The Carley Barn, raised by hand at Churchfield Farm for Esther and Bruce Carley, is the result of Barn Club's efforts. Woven into the story of how the Carley Barn was made is an essential message of handcraftsmanship, intimacy with nature and the companionship of the volunteers. More and more of us are recognising the limitations of our modern life and yearn for something better, more meaningful, honest and fulfilling. We yearn for ways of life that make better use of the planet's resources and that meet our human needs. Access to nature is a big part of this, and helps us to find routines that nurture a deep personal relationship with the Earth.

Where do we start? At least to begin with, we could learn a lot from the past by rediscovering, if we can, the things that have been forgotten. I have no doubt that some of the changes that we seek to make in our modern world were once completely entwined with daily life for previous generations, from a time when nearly everyone was a maker using their hands for their livelihoods.

This book explores the nature of rural craftsmanship and follows the process of making a barn in the way it used to be done. My hope is that it will reveal some of the ingredients for a way of life that was so taken for granted at the time that it was, for the most part, unrecorded in history books. However, we do still have the evidence of these makers' handiwork, not least the timber frames of historic agricultural buildings.

The choice of elm wood as the structural material for the Carley Barn, and the extensive coverage of this tree species within this book is, of course, intentional. Why was elm wood used and not oak? What is it about the tree that made it so important?

In the century before the Industrial Revolution elm was the timber of choice for structural frames in the region where I live as well as for countless household artefacts. Elm is a species that suffered a major pandemic, but its incredible determination to survive prevails. Elm is proving itself to be a tree with an enduring life force and, to my mind, is an appropriate icon for getting closer to nature, the resurgence in making things by hand and for bringing old skills back to life.

Elm is one of the least understood trees in Britain. As yet, there is no national public body dedicated to researching these trees and to restoring them to the landscape. Elms have become the forgotten tree, the tree we gave up on. Most people under fifty years old will not have even stood under a mature elm and marvelled as they looked upwards. Elms are likewise not currently given a second thought in tree-planting programmes.

Fair enough. It's understandable. To value and grant importance to something it helps if you have experienced it yourself. Yet, unbelievably large elms do still thrive and are dotted about across Britain. It is in our generation's hands to undo the wrong that we unwittingly did when we imported a deadly disease to this tree in the twentieth century. I think the first step is to find, stand under and cherish the elm trees that still live. And when they die, as all trees must, use the timber. It is a joy, as elm is unlike any other native wood. For modern society the elm is a tree to be rediscovered. Then we can collectively resolve to instigate and celebrate their re-emergence in great numbers as healthy trees in the landscape. This will not be directly for our benefit, but for future generations of people and also for wildlife in Britain. From what I have learnt and observed, elm trees and their occasional hollow cavities provide homes and habitats for creatures of all descriptions, not least wild honeybees. The bee and the elm. Perhaps these two species are more linked than we have realised, and together they can be brought back from the brink.

Inspired by the sight of the Great Barn at Wallington, an ancient elm barn on a farm near to my home, I set out to follow the process of making that had been undertaken by people of a previous era; before the days of power tools and mass manufacture. I wanted to follow their footsteps. I wanted to learn firsthand why they chose elm wood rather than the customary oak. My vision was to incorporate as much as possible of the process that those carpenters undertook to be able to understand their challenges and their joys. From the beginning of this project, friends, neighbours and even people I had never met before were attracted to the idea, and offered to come and help for free. Barn Club was born out of this spontaneous, shared willingness to engage with my vision and be part of the event. The emergence of this sort of community spirit has been both heart-warming and humbling. In my life as a designer and carpenter, I have always assumed that you needed to pay people to do stuff, whereas the volunteers see things differently. Why would anyone choose to come and work for nothing? It's a question I kept asking myself. Adrian Toole, a maverick cyclist, enthusiastic green campaigner and also someone who saw the project from start to finish, commented:

> *My mother could not understand it. 'Why do you keep going when he doesn't pay you?' she asked time and again. The answer was that I knew those days working with the wood were unique life experiences and I appreciated the laughter, the teamwork, the training and, of course, the lunches!*

The volunteers made their livings across a broad range of professions, from accountancy to bicycle safety instruction. For many there was a strong desire to be doing something other than sit at a desk in front of a computer; Barn Club is a really good relief from digital overload. Making something beautiful is also massively satisfying. Most importantly, timber framing done in a traditional way is accessible to everyone. The volunteers were able to fully take part with no prior carpentry knowledge. They picked up the skills whilst using hand tools just as novices would have done in previous times. Nick Exley, a retired dentist, with a wind-blown fitness and a love of the outdoors, describes his experience:

> *Once I had completed Robert's carpentry course, that initiation into the 'mysterious arts' of timber framing, I very quickly began to look forward to my days working on the project. I*

*have always enjoyed working in small teams and this didn't disappoint. It was happy and without pressure because of the concept of being volunteers and not on a strict deadline to get the job done. There were moments of bafflement when we would be working in pairs and that unspoken glance would say, 'Do you understand this?' And the look back said, 'No, I don't. Do you?' Even in the showery weather we could huddle in the lean-to canteen for a tea break and put the world to rights. But somehow we always got there in the end, and as each week went by there slowly emerged possibly the most beautiful building I have ever seen. But of course, that's because I could go round and say, 'I did that bit!'*

There is no doubt at all in my mind that resisting the lure of power tools was key to finding new possibilities for working with volunteers, and therefore a new approach to working within contemporary woodwork culture. Volunteers need to work in a safe environment, and power tools are essentially unsafe. To be able to work with inexperienced people, hand tools are the answer. As Joel Hendry, a leading timber frame carpenter, puts it, 'When you start using a machine morticer, you pass a threshold. Before long the way you work, the type of joints that you make and the whole circumstance of the framing yard changes to suit the machine's needs. In the worst case, a carpenter can become little more than an attachment to a power tool.' When I asked Joel why he liked timber framing by hand he replied, 'The joy of using hand tools is that it needs hand, eye and brain, which always feels good.' Joel isn't an advocate of a power-tool-free world, but by having a foot in each camp he acknowledges that there is a price to pay when machines predominate. As a highly skilled individual, he abhors the treadmill of nonstop, endless, frame factory production using power equipment. Why would anyone volunteer to do that?

The revival of timber framing in England is ongoing and expanding, but that hasn't always been the case. By the early twentieth century, the scribing and cutting of local timbers, then fixing them together with wooden pegs, had nearly vanished as a way of building. Along with this decline went a loss of knowledge. Two notable exceptions are the timber frame for Bedales school Memorial Library by Ernest Gimson and Geoffrey Lupton in 1921, and the new hammer beam roof of the Great Hall at Dartington Hall estate, by William Weir and Jack Goode in 1933. These two buildings stand out not just for being large timber frame projects based on what had by then become bygone traditions, but because the architects themselves were working craftsmen.

The current revival began about forty years ago, and the Carpenters' Fellowship was established in the UK at the beginning of the twenty-first century. There are now many companies and individual framers spread across the country producing timber frames for houses, garden studios and wedding venues, and where I live there is a thriving fashion for porches to embellish front doorways. The demand for chunky, bare oak frames with an aura of craftsmanship hasn't diminished in our modern era. At the same time as the appeal of post-and-beam carpentry has expanded, the physical process of producing the frames is increasingly mechanised, efficient and automated. In an age of digital industrialism, this is a good moment to review what is at the core of the craft to ensure that the important things don't get eclipsed.

My memories of the early days of the timber frame revival in England are of a shared passion for wood, for good-quality tools and for the discovery of an obscure form of carpentry.

And, most importantly, I remember the fun of getting together and making a project into a joyful social gathering. Carpenters would travel around the country to work on different frames and for some particular projects a gathering would be arranged. These became affectionately known as 'rendezvous' projects, reflecting the spirit of camaraderie among timber framers. Volunteer carpenters would arrive en masse and stay for a week or so until the frame was completed. This spirit lives on today, and rendezvous projects occur across Europe and North America, particularly through the efforts of the Carpenters' Fellowship in the UK, the Timber Framers Guild in the US and an international group called Charpentiers sans Frontières (Carpenters without Borders) based in France.

Although not unique to timber framing, the atmosphere of generosity and goodwill are something that surrounds the craft. In the late 1990s I travelled across the US and found the same thing happening there. I was warmly received by timber framers and their families, from architect and builder Jack Sobon on the East Coast to the late Merle Adams of Big Timberworks in the Northwest. Those weeks spent observing their different business models and experiencing different rural cultures and carpentry traditions inspired me deeply and changed the course of my career. I wasn't travelling as a working journeyman, though those ancient traditions of medieval guild apprenticeships still survive in continental Europe today. Apprentices still travel between carpentry businesses as part of the training, carrying only their tools and often wearing traditional flared black corduroy trousers. This aspect of travel and exposure to diverse cultural experiences is thought to be an essential part of what it means to be a *compagnon du devoir* – companion of duty – within the French system of travelling craft apprentices. Learning a craft involves so much more than acquiring a skill or operating machinery to produce a product. In the past this was not only understood, it was fundamental. The *compagnon* would be on a personal journey of self-discovery that had moral dimensions as well. Learning a craft was meant to be a life-changing experience.

Mihai Vatajelu, a timber framer who runs a construction company in London, has taken part in the gatherings of the Charpentiers sans Frontières and came to volunteer at Churchfield Farm. Like so many people I talk to, he has grandparents or great-grandparents who were carpenters. He also feels the tug of wanting to work in a more meaningful way with wood:

> *I love timber framing. My best memories of the Carley Barn are the mornings at the start of the project, especially the cold ones; light snow on the timber, and then warming up through the physical effort of the carpentry. Working with wood the traditional way seemed to trigger in me a sleeping gene that, as a Romanian, I'm sure must have been passed on to me from my grandfather, Simon Vatajelu. He was a forester in charge of a large birch tree forest. I'm trying to find a way that makes sense in combining the new way of life dominated by technology with the traditional way. Always choosing technology will only make the old ways disappear.*

I believe that when we think of making something and we set out to gather the materials and the wherewithal to do it, we stand at a fork in the road. On one side we see a well-trodden path of power tools and technology, on the other, a path of handwork, far less travelled in our day and age. By taking this second path, as we did to make the Carley Barn, a different world opened up.

Immediately, in the absence of machine noise, you can hear what you are doing as well as see it. You can hear background birdsong or the soughing of the wind in the branches of trees overhead as you work. If you are able to source your raw material from your immediate natural surroundings, as a woodworker can, knowledge of the natural world seeps into you unconsciously. This connection and dependence on nature changes how you live. A craft that is rooted in its materials inevitably places you within the natural world rather than being simply an observer of nature. You develop a personal relationship with the landscape and the lives of animals, insects, birds and plants within it. This reaching out into the natural world to find the essentials of your craft occasionally leads to feelings of sublime happiness. It is not surprising really, as every single cell in our body is part of nature and is completely at home when we walk through a wood, stare at the sky or watch the flowing of a river. We have subliminal capacity in these settings, which fires up particular intuitive knowing. We begin to feel very much at ease and at one with the landscape. An uncanny knack of finding what you are searching for develops. In my case it appears through love of trees. Finding elms seems second nature.

Further along this 'less well-trodden path' a sort of intimacy grows, which is felt through the close partnership of hand tools, our hands and the material itself. Sometimes when you are working, the distinctions and boundaries between these three break down and you lose all sense of which is guiding and which is following: something is simply appearing through itself. This level of surrender to the process has elegance and grace, which I feel certain is a shared experience amongst many crafts. A craft is more than a hobby or a trade, it is a way of giving life to things.

There is an additional satisfaction when working together with other volunteers. A bond develops between people born of goodwill, without the need for an economic contract. This became very apparent over the months of making the Carley Barn. Coupled with the sharing and learning of a particular craft tradition, great companionship was forged. This is what is hinted at in the phrase 'intangible cultural heritage' coined by the United Nations Educational, Scientific and Cultural Organization (UNESCO) to describe the particular practices of a community that are handed down through the generations. It is more than just a craft tradition; it is something that is experienced as a drawing together of people. When we come together to work in this way, something comes to life that also goes beyond the boundaries of time. We enact a craft in the moment, in the present, but we do so as agents for a tradition that goes back in a chain of companionship with our ancestors. It feels a very natural and human thing.

There is without doubt a desire in modern society to find the time to make things ourselves, with our own hands. We are attracted to meaningful experiences – the process, not just results. People don't just like a beautiful garden, they like gardening, don't they? The volunteers had different reasons for being involved, but the love of being outdoors and making things was a common thread. Brian Bernard, a retired engineer and a man of dogged determination and an eye for detail, describes his attraction to the project:

> *I volunteered for the barn because I have enjoyed working with wood ever since woodwork O Level at school. I also prefer to be outdoors. I remember going camping and making fires with the Scouts, and was fascinated by the* Out of Town *programme on TV in the 1960s when Jack*

*Hargreaves would explain different woodland crafts each week. That's what drew me to Barn Club.*

Currently there is a revival of making and mending. It feels fresh and exciting. In some ways this emerging culture is part of an ongoing craft revolution that threads back in time through people like Jack Hargreaves. I think it's a continuing process of something appearing that was always there. Something that was lost or forgotten, and that, actually, we find quite amazing and relevant today. Like discovering an old hand tool, say a bruzz, that you haven't even heard of before and finding that it is absolutely the perfect tool for a particular job. Or learning to trust your hand and eye as you saw a kerf to a line. One volunteer who came to the barn raising commented that he had been told he was no good at carpentry when he was at school and so hadn't bothered with it since. I wasn't going to stand for that, so I invited him to help by cutting a tenon for the end of a floor joist that was being measured and fitted in situ. With a little coaching he was able to cut a perfectly straight line with a handsaw, overcoming the inhibitions of a lifetime. He discovered a new skill he didn't know he possessed: 'I was so pleased with my one tenon, so all the others who have contributed so much must be feeling incredibly proud. The craftsmanship to make sure that every joint fitted is awesome.'

Today we are in a process of rediscovery of things we did not know existed or were even possible: the extraordinary richness of the natural world on our doorsteps, the brilliance of simple hand tools and our own latent dexterous capability.

The satisfactions and benefits to be found in something like Barn Club are undeniable. Furthermore, the environmental impact is benign, even positive: trees absorb carbon dioxide and convert it into organic chemical compounds that form their bulk and structure. The locking up of several tons of $CO_2$ in timber for potentially hundreds of years makes timber framing by hand, in the way described in this book, a carbon-capturing activity. By using the timber of dead trees, or trees that have already been felled for firewood as I often do, the capture of $CO_2$ is maximised. Even the careful selection and felling of mature trees, if conducted with restraint, replicates the natural cycle of life and death of virgin forest. Nowadays, the good management of deciduous woodland also includes intermittent felling of tightly spaced, mis-shaped or injured trees to allow light and air to those that remain and to the saplings beneath as well as the flora of the woodland floor. Woodland biodiversity expands when the tight canopy of branches and leaves is opened up a little in this way, creating glades and pools of sunlight. These timber thinnings are normally burnt as forest waste biomass in contemporary forestry but are still valuable and usable to a carpenter earning a living through traditional crafts. If we are ever to tackle the climate crisis, we need to explore all the ways to build structures without generating more atmospheric pollution. I have calculated that the traditional methods described in this book results in a tenfold increase in the amount of useful timber from our local landscapes, without threatening their existence. Is Barn Club a model for further projects elsewhere to be taken up? I hope so, and this book might help. But I am certain that a 'new wood culture' is possible. Without contradicting our love of trees, it is possible to plant, nurture and care for our local deciduous woodlands with a useful product in mind, so they can provide us with beautiful timber to make long-lasting structures. And all done in ways that promote companionship and well-being.

## TWO

# Starting Points

There is an ancient elm barn in Wallington, Hertfordshire, not far from where I live. It is the barn at the heart of the story *Animal Farm*, written by George Orwell and published in 1945. I had come across the barn at the invitation of the owners, Nick and Diana Collingridge. They mentioned that George Orwell had lived in Wallington in the 1930s and that Manor Farm and its Great Barn were the setting for his book. As a child I had seen the Walt Disney animation of the story that depicts the farm and the barn, with the final, single rule painted in large white letters… 'All animals are equal, but some animals are more equal than others.' So one spring day I set out with Nick and Diana, who proudly showed me round the building that is now a place for the celebration of village events. The exterior is unremarkable, just black-painted horizontal boards and a dark slated roof. But the inside is another world. This is where the inspiration for Barn Club began.

To step inside this barn, into the darkness within its walls, is to take a step back in time. Not just to the 1930s, but far beyond, to an indeterminate, distant age. The barn is dark inside, almost black. Your eyes cannot make anything out. It is cool, still and silent, except for the slight rustle of the breeze over the slated roof high above. The smell is not of wood but of straw – old straw. This was once a threshing barn, but now has not an ear of wheat inside. The smell lingering in the air comes from countless harvests that crossed the threshold every autumn: the harvests of generations of farmers. Once inside it takes a while for your eyes to accustom to the gloom and the shape of the light switch next to you to reveal itself. Slowly, the old sodium lights warm up. But in those moments of semi-darkness you feel the atmosphere of the barn. Your eyes begin to distinguish the great supporting posts around you. You can almost imagine you are in a glade surrounded by tall trees during a twilight walk. In these enormous barns you still feel the presence of trees. All around is an array of timbers leading up and out from the posts, diminishing and arching as they go, like the branches and branchlets of a tree. High overhead, way up, are thin battens, the twigs, supporting the roof slates. The outer walls are completely wrapped with timber cladding to make an enormous wooden box that glows a subtle honey colour. When the lights have fully warmed you also see ahead a worn, gently undulating floor, crazed with cracks from wear over the decades. Entering the towering space feels both uplifting and humbling. Like many old barns there is a silence and peace within the encompassing structure. It has an undeniable beauty. People say that these barns are cathedral-like spaces. They are. They are like cathedrals to nature.

The Great Barn at Wallington was built over two hundred years ago, raised upright by human effort. The mortice and tenon joints were cut by hand, and the timber came from local woodland and hedgerows. No architects or engineers were involved in its design, it was made by common carpenters and farm labourers.

It has various Roman numerals, initials and dates scratched into the timbers inside, and, high up in the roof, one inscription reads, 'Built E.F. Farmer 1786'. Another, on a post, reads, 'G + C built 1786'. The lines of central posts are set out in a precise form in the pattern of the local carpentry tradition, well understood at the time and handed down from father to son. The barn is divided into four bays by the posts, with double cart doors in the front. There are two aisles either side of the long, central area. It is easily big enough to hold an entire harvest of wheat or, equally, a wedding feast.

It is still used by the village for local events and at just such a barn dance I met an elderly local who could remember a carpenters' workshop in the nearby town of Royston. The company was called Farmer and Sons. The business is no longer there, but in 1976 the carpenters made half barrels as flower planters for the town to celebrate the jubilee of Queen Elizabeth II. It is tantalising to think that they were the same family of carpenters that had cut and raised the barn at Manor Farm. However, since the person who paid for the barn was a farmer named Edward Fossey, it is more likely that he is the 'E.F. Farmer' in the inscription, and that 'G+C' refers to the carpenters. I think that perhaps the latter is true. Most barns, being practical vernacular structures, carry no inscription at all. They were built by unknown craftsmen.

Inscribing a timber with the date and their initials might occasionally be a final act of completion for the carpenter, but there are always other marks as well. Each piece of wood has a Roman numeral to locate it in the greater scheme of the construction, and there were indeed carpenter's marks in this timber frame. Nick produced a torch so that we could shine a light up along the length of each piece of wood to highlight any more detail. There were the marks of pit saws and side axes as well as the just discernable thin lines of a carpenter's scratch awl used to set out the timber joints. It feels very special to inspect these marks made over two hundred years beforehand. There is a kind of intimacy and linking of minds across the centuries as you closely observe these traces of the original carpenters' work, perhaps the scratch of an awl made with the flick of the wrist in a fraction of a second all those years ago. Close examination of the heads of the main posts also revealed a curious feature. The posts widened at their heads to form projecting jowls and there were carved profiled shapes, about four inches long, at the chins of these jowls. See top left image on page 17. Within the whole of the structure, these jowl chins were rare places of deliberate, non-functional decoration.

A traditionally constructed aisled barn like this one, built at the end of the eighteenth century, is in itself unusual. It would have been made with materials and methods developed over millennia by local craftspeople in a time-honoured way, accepted and understood as the way things were done. But it was also created just a few years before a wholesale transformation of the practice of village carpentry. At about this date there was a shift to using imported conifers, machine-sawn timber, engineered truss design and metal fixing methods. Even the way of sawing wood was changing. The circular saw was 'An entirely new machine for more expeditious sawing...' patented in 1777 by Samuel Miller of Southampton. In 1799 Marc Isambard Brunel built the first steam sawmill at the Chatham Dockyard. Not long afterwards, in 1808, William Newberry of London invented the band saw.

The Wallington barn is one of the last of its kind in the region. I can imagine the head carpenter being a man who wanted to show the apprentices how things were done when

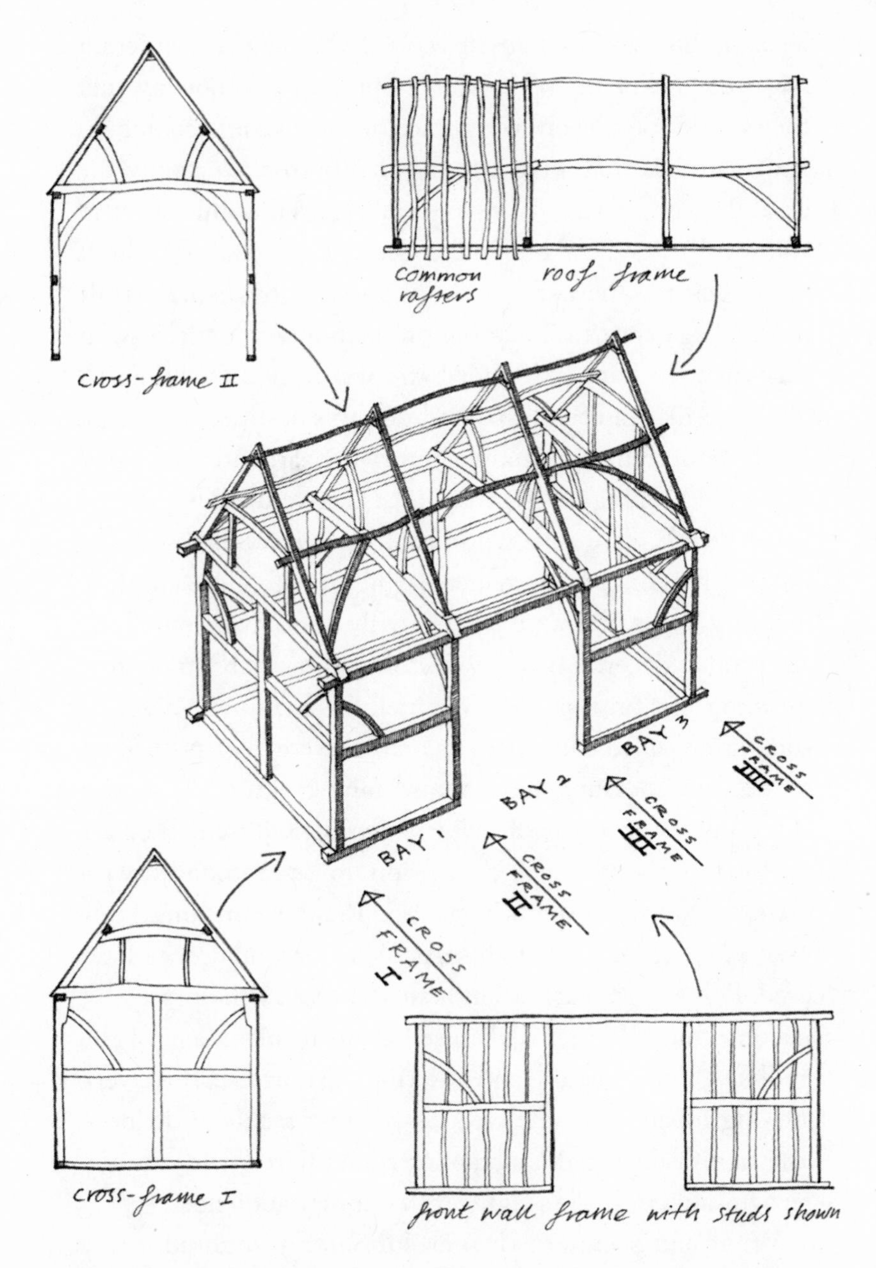


The bays, walls and roof of a typical Hertfordshire timber framed barn after Richard Harris.

he learnt his trade. Perhaps he wanted to show all the different styles of jowl chins to the young apprentices, knowing that they would soon be off to embrace the new styles, techniques and materials that were arriving in Royston. Certainly the great thatched barn at Wimpole Hall, just a few miles away in Cambridgeshire, and built a few decades later, would be built with an entirely new and modern approach. By then, the shift from village craft to industrial production was underway. In this sense the barn at Manor Farm was itself a final flourish of an age-old tradition. This tradition was destined to die out. By the end of the nineteenth century the carpenters who had made the Wallington barn would also have died. Fewer and fewer people were alive who could explain how these magnificent buildings were made, leaving only folklore and folk memory. Tools that were once in daily use can sometimes be seen today decorating the walls of country pubs; their true meaning lost on a modern clientele. For example, dividers, sometimes called compasses or 'points', were universally used to transfer measurements before tape measures. The two sharp points of a pair of dividers can be adjusted. They are set to record a particular dimension, to be reproduced with extreme accuracy somewhere else. They are an amazingly useful tool and one that all sorts of trades would once have used. Points were such a fundamental tool in the process of making that William Blake used them in his image of the godlike Urizen creating and ordering the universe. It is a very striking image, with the creator wielding a pair of dividers. The significance of the metaphor would have been instantly recognisable to Blake's eighteenth-century audience.

What had been a gradual evolution of a localised, orally transferred knowledge came to an end at about the time this barn was raised. This knowledge was lost over the ensuing hundred years. Obviously barns were still used and repaired as needed, but as the knowledge of framing carpentry thinned, the repairs became ever more crude. The evidence of local barns in Hertfordshire also catalogues a switch to stud frames, fixed together with nails, straps and bolts rather than wooden joints. That is not to say that there is no skill involved, but aspects of the old craft were being lost along the way. The switch to imported conifer softwood also meant an influx of straight, square and stable material that was easy and quick to work with. At the same time, the increasing invention of machinery and powered equipment for almost every stage of the process, from a tree to a building, led to a downgrading of human involvement to the point that it can now, through automation, be sidelined almost completely.

The question is, what might have been lost? What quality of living well, in harmony with nature, have we forgotten? What can be discovered through the ways that things used to be done, and might still be applied today? What can the Great Barn at Wallington tell us about the process of making that we might still choose to embrace?

Having closely inspected some of the timber of the Wallington barn, I turned to Nick and said, 'You realise this barn is made of elm?' I was surprised by this discovery, but not as surprised as Nick, who had assumed it was all made of oak. I had occasionally seen elm used in houses and barns before, such as in the attics and barn roofs at Kelmscott Manor in Oxfordshire. But I hadn't yet found a complete structural frame of elm. The fine zig-zag lines in the pattern of the grain indicated elm, but, standing back, I could recognise the soft, creamed-honey colour of old elm wood. It is subtly different from oak structures, which are more straw coloured in tone.

It was a real treat to find so much elm wood in one place. Elm trees suffered a massive decline in the twentieth century across Western Europe and North America, as a result of a non-native fungus that killed 99 per cent of mature elm trees and is now endemic in our landscapes. Both the elm tree and the elm timber, like the points that William Blake painted, were once ubiquitous, but now have dropped from public consciousness.

The English elm, the really tall type of elm with a billowing outline, once dominated the English landscape, rising higher than oak trees. It is very hard to imagine, almost impossible, unless you were alive at the time. In some regions, elm trees were more numerous than oak trees are today. In fact, one of the earliest photographs ever taken was a silver salt print 'calotype' of an English elm at Lacock Abbey in Wiltshire, taken by William Fox Talbot in circa 1843–45. It is an iconic photograph, not just because it heralded a new age of photographic printing using silver salts, but as a record of an archetypal tree of the English countryside: an English elm over 100 feet tall. The photograph is taken in winter, with the structure of the tree on full display, and subtle clusters of blossom just discernable on its topmost twigs. The trees in the background are young elms, more typical of the extent of the growth of English elms today. If they could survive to maturity, they would become magnificent trees like the one in front. This particular elm tree no longer exists, but at the time it was famous enough to be called The Great Elm. It had survived the storms that had brought down many other large local trees in the preceding years. This tree shows no sign of damage. It is a magnificent tree that obviously meant a lot to William Fox Talbot.

Along with oak and ash trees, elm grew in almost every hedgerow and village in Southern England. It had been deliberately planted during the seventeenth and eighteenth centuries as a timber tree in boundary hedges. In those days the total number of mature trees existing in hedges and on farmland exceeded the total of woodland trees, which is an extraordinary thought. These elms were often planted by owners of large estates as a result of enclosure of the commons and as a tenant obligation placed upon farmers. So, the late eighteenth-century barn I was standing in could well have been made from some of these deliberately planted hedgerow trees.

If the Great Barn at Wallington had inspired the story of *Animal Farm,* it was also the barn that pricked my curiosity to think about forgotten skills, forgotten carpenters and the forgotten elm trees. What else might I be overlooking? Standing in an ancient barn made of wood, you see the beautiful finished product, but you don't see the people that made it or the way that they made it. The makers' faces and their hands are invisible. By focusing on the product, you don't see the process. The barn is an undeniably beautiful space. Was that beauty deliberate and conscious? Or was it due to the process? Modern farm buildings are so lifeless, almost brutal by comparison.

We now know fairly well the types of wooden joints that were once used and how to make them today. They have been named and classified. But a barn is more than an arrangement of timbers and connections. What did the process of making a barn from scratch actually feel like? That inspiring visit with Nick and Diana was the catalyst that formed these questions.

The following year I set out to discover and sketch other local barns, and I soon realised that there was a regional tradition of using elm in East Anglia, particularly in the eighteenth century. I also spent numerous days out searching for elm trees, with unexpected success. So far this has resulted in two new elm barns being built by local people from local trees. Perhaps these are the first barns to have been hand-built in this way for two hundred years. This amounts to a sort of renaissance of a local craft tradition, though it is much more than that. Barn Club has been a focus to renew our love for trees, for being out of doors, working with wood, using hand tools, and the rediscovery of the companionship and possibilities of teamwork.

By focusing on the process, a barn can be seen as an event as much as a physical thing. A fellow carpenter, Kris Vill, said, 'You aren't just building a barn, Robert, you are building a community.' That dedicated teamwork and a willing community could be integral to a craft had not occurred to me before. If it takes a village to raise a child, perhaps it takes a community to build a barn.

It is difficult to give a name to this aspect of a craft. Today, the word 'craft' has evolved into wide-ranging meanings. It is applied to so many things, from writing a book to roasting coffee. Common to most concepts of craft would be an indication of great care being taken and the close attention of some human faculty. We see contemporary crafts as creative, individualistic and a means to produce objects of aesthetic value. However, I don't think we have a simple word in English to identify the approaches to making something in ways that are the subject of this book.

The word 'craft' has evolved in its meaning over time, and continues to evolve. In the UK it was conceived in its modern sense at the end of the nineteenth century. It arrived with a great fanfare. In 1887 a group of British architects, designers and makers decided to form the Arts and Crafts Exhibition Society, the name coined by the bookbinder, T.J.

Cobden-Sanderson. As the name suggests, it was set up to organise exhibitions to promote the work of designers and makers in applied design and handicraft. They believed that artists who made and decorated physical objects should be valued as much as painters who hung their finished works on gallery walls. There were many creative people involved in what was as much a social movement as a purely artistic one. Although inspired by the progressive philosophical ideas of John Ruskin, one name that stands out today is William Morris. He famously said, 'Have nothing in your houses that you do not know to be useful or believe to be beautiful.'

Interestingly, there was also a much less famous and almost underground architectural movement that was inspired by John Ruskin and William Morris. The philosophy of the 'wandering architects' in the UK survived up to the First World War through the work of people like Detmar Blow. The revolutionary idea was that architects could also be fully competent craftspeople as well as draughtsmen and would build their own designs, often spending many months living on-site.

Barns, as functional buildings with very little applied decoration, would have had less appeal than houses to the original patrons of the Arts and Crafts movement, other than their inherent honesty and rustic simplicity. Carpentry as a rural craft was practised at that time by unknown, unnamed individuals, rather than the designer-makers that were making a name for themselves, following William Morris's inspiration. The Great Barn at Wallington was built exactly 101 years before the phrase Arts and Crafts was invented, so perhaps the Arts and Crafts movement is not the right context in which to consider the carpentry we were following. The word 'carpentry' means something very different today. It is still about working with wood, but in the past there were other important connotations that we don't even imagine today when we use the word.

From accounts of life as a carpenter the late nineteenth century and early twentieth century by the English writers Walter Rose and George Sturt, there was a strong connection between carpenters and their local trees and countryside. Writing about village carpentry of his grandfather's days in the early nineteenth century, Walter Rose wrote, 'The country carpenter looked to the trees growing nearby to see how each could be used for the work to be done.' The local supply of materials, knowledge of how trees grow, how to move the heavy tree trunks by hand, and how the wood is sawn and seasoned makes comparison with a modern workshop very stark. This knowledge anchors the craft traditions to the locality and gives a sense of place to the woodwork. Although it is not explicitly stated by these authors, this is the context in which carpenters knew their craft.

In pre-industrial times there would thus have been a close relationship and often a physical proximity to the source of craft materials, which were experienced as being found in the natural world, whether it be the trees from the forest or the sand from the river. There was an inherent provenance to every product. Modern materials have no readable story, and their supply chain is anonymous. Materials have become commodities, and without a narrative they are lifeless. Like a sheet of plywood or a bag of builders sand. Yet, in reality, no two trees are exactly identical and each river bed subtly unique. Each has a different story that is written over many decades or centuries. When the timber is sawn or cleft, the story is revealed. It is in the grain, in the physical objects that are sometimes discovered embedded deep within the timber and also in the memories of the people who lived amongst the trees as they grew.

I love being a carpenter working out of doors. It is a privilege to visit the wood with the clients and volunteers to find the trees that will become the timber frame. Staring up through leaves growing 80 feet above, we can share the sense of wonder at the enormous beings around us. When the carpentry is finally underway, because we are doing it outside in a yard, we also experience the turning of the seasons and the events in the landscape each day. Whilst working on the Carley Barn one afternoon in late May, concentrating hard on the woodwork, we all lifted our heads to watch a swarm of honey bees approaching over a field and passing nearly overhead. The sight could have been played out two hundred years beforehand, experienced by the carpenters of the Wallington barn. The bees moved in an incredible, swirling cloud and with a roaring sound that has to be heard to be believed. They all circled at crazy speeds, thousands together. Though the cloud itself, much bigger than I had imagined, moved away at a dignified pace. We all became dumbstruck. Witnessing a bee swarm is an experience that everyone should be able to have once in their life, if they can. I had to wait for fifty-seven years before I would see it with my own eyes.

For the most part, though, being outdoors is an experience of stillness and quiet. You realise that we, the humans, are the noisy ones. Some days in the framing yard at Churchfield Farm there was just the sound of the breeze in the tall lime trees and the cackling of jackdaws. At times nothing seemed to be happening. Or, at least that is how it appeared. In the same month as the bee swarm, Esther Carley showed me some bee orchids that had appeared next to the large pond where we were working. They are small and easily overlooked; its flowers the size of a bee, as you would expect. But when you look closely, their beauty and intricacy is astounding. It's all the more remarkable because you know they are a simple, unconscious flower beautifully formed in detail like a bumblebee. They seem incredibly special and worthy of a special journey to observe, yet there they were, growing just yards from where we had been working all spring. I can only speculate at the experiences of people in the past, like those who made the barn at Wallington, working outside when birds, insects, flowers and wild animals were all much more numerous. But at least Barn Club allowed us every opportunity to stop and marvel at the natural world. Rob Mills, a commercial accountant with a thoughtful, gentle nature and a ready smile, experienced this whilst making wooden pegs to be used in the barn:

> *I was filled with calmness, sitting on the shave horse whilst making my first peg. I surveyed the scene across the pond, spotting the purple of the bee orchids and gazing on to the spreading limbs of the great cedar tree in the distance.*

Nowadays the word 'craft', particularly carpentry, presupposes activity indoors in a workshop and in reality its practice relies on machine use and an increasing digital automation that William Morris and those before him would not recognise. There is now a near universal acceptance of power tools or machine assistance among contemporary crafts people. But when trying to make a barn in the way of our forebears, power tools would have to be restricted. If you have already used power tools for many years, it takes great effort to resist reaching inside your truck for a circular saw when confronted with the task of cutting through a piece of 6″ × 4″ timber. I feel this tug continually. I also know the difference in how I feel after a day of work using hand tools and a day using power tools. The latter is not only exhausting, but dangerous as well. Unsurprisingly, you wear protective goggles, masks, gloves and earmuffs, reducing peripheral awareness whilst focusing all your attention onto the machine's manically spinning and lethal cutting edges. Conversely, when you sand a piece of wood by hand, no protective equipment is needed, and it is immeasurably more satisfying than using a powered machine. It is worth taking the time to try it – if only once. Working by hand you experience the gradual appearance in front of you of the detail and beauty of the grain as something you are personally drawing out from the wood, rather than trying to rush through a necessary but noisy and dusty job. Stranger still, it seems to take less physical energy when done by hand.

Machines demand you work at the speed and requirements of the machine. This prevents you working at a pace and in a way that is in tune with the task at hand. In his book *The Wheelwright's Shop*, George Sturt reflects on the first introduction of a machine lathe and saw bench into his woodwork shop in Surrey in the 1920s. He makes the surprising observation that his men had more physical energy after working fourteen hours each day with hand tools than they did after twelve hours of using machinery. Productivity increases with machines, but the work can be more tiring. The noise of a modern workshop is formidable. Circular saws, impact drivers and nails guns are incredibly loud. In contrast, the swish of several handsaws is a delightful sound, and you are able to chat as you work. It sounds a simple thing, but being able to talk makes a big difference to well-being, and the ability or even willingness to co-operate with each other. Hand tools, well made and kept sharp, are a pleasure to use. They can last a lifetime, some of them even longer.

After a while, their use becomes second nature, graceful and intuitive. You get to learn to feel the tool's edge as it interacts with the wood. When you have had enough practice it becomes effortless. Some tools can even be made from what you can fashion from your immediate surroundings, like a mallet made from an overgrown hedgerow holly.

So, in this way of making there is an immersion in nature, a joy of hand work, and also a third part that is a vital component and fundamental to Barn Club: that of working together. Of all the possible evocations of teamwork, the raising of a timber frame by hand is one that is most visually resonant. The satisfaction of seeing the frame appear, using no more than combined human strength is awe-inspiring. Seeing the hundreds of lengths of wood all come together in the course of a day and match perfectly is really impressive. For many weeks and months leading up to the raising, the people of Barn Club gathered to work on the frame, one piece of wood at a time. Working together developed into a companionship and satisfaction that I have never experienced when working alone or with machine tools.

Our ancestors knew this well. They would organise and co-operate as a means to get things done, before the advent of cheap energy and labour-saving machines. One volunteer, Vince Kilcoyne, a banker specialising in financial technology, whose silver Irish wit and bounding imagination always made me smile, grew up on Achill Island, off the west coast of Ireland. He remembers going up onto the bog with fellow villagers to cut the turf, the peat, that would be stacked and dried for each of the cottages as they needed it. He said it would be a communal event; going up to the common together. No payments were made for the task. It just needed enough hands to make light work of it.

My family did something similar in the 1970s. When I was a child, we would stay in Devon in the summer to help with the hay making on a friend's farm. It was in the days when bales of hay were small enough to be lifted by hand. We walked up from the farmhouse, up the hill behind and past sweet-smelling hedgerows to the hay field, returning in the evening riding on the hay bales on the back of the trailer. We all enjoyed it. It felt good to be part of the farming year.

Similarly, my mother grew up on a farm in Scotland. She remembers when she was a girl helping to stook the wheat sheaves in the field to dry, and in the evening when the horse-drawn wagon was piled high with dry sheaves lying on her back on top, being brushed by the leaves of the trees in the lanes. Large numbers of people were needed for harvest. A harvest was a significant event in rural life requiring many hands, with a feast laid on for the dozens of people who had helped. Even the village carpenters would stop what they were doing to go and help gather the harvest, as Walter Rose relates in his book, *The Village Carpenter*. By comparison, harvesting the wheat in Hertfordshire today can almost be done by one man alone, switching between the combine and the tractor. It takes our neighbour just a day or two to harvest his 20-acre field next door.

Not only would large numbers of people have gathered to undertake collective tasks, but they would also have had the knowledge and skills to be useful. In the South West of England wheat straw was traditionally used for thatching. The piles and stacks of items that today we cover with a tarpaulin would have been given a waterproof thatched roof. Even the stacks of kindling were thatched. Everyone learnt to thatch. The ability to thatch was shared by men, women and children, not just the thatcher. Handcraftsmanship was part of 'community capital' in rural society.

This can all seem romanticised to modern eyes, and I am sure it was hard work if there were not enough people. But perhaps we romanticise the things we really miss and yearn for. The possibility for communal physical activity is still available to us today, as demonstrated by the growing movement for communities to plant trees. I recently joined in on a planting. There were about fifty people. We planted a few hundred trees together for a Devon farmer without breaking into a sweat. It was fun and really easy. Looking back over the freshly planted wood, it seemed staggering that so much could be done so easily, rather if the farmer had planted the trees on his own. The experience of working together, chatting as you go, and having a meal with the people you work with, forms great bonds of companionship. To become a valued part of a team is a healthy place to be. There is a feeling of shared achievement and pride at the end of each day.

Goodwill is at the core of this. It would be true to say that goodwill also goes a long way to building a barn. After all, Barn Club consists of volunteers. As one such volunteer, Greg Cumbers, put it: 'When you arrive, there is no money on the table. Money doesn't come into it. You can leave when you like. You are there because you want to be.'

Goodwill and co-operation are priceless and together create a happy working environment. Co-operation allows people of all sizes and physical ability to be involved. Counter-intuitively, you don't need to be abnormally physically strong to be a carpenter. Most of the timbers in a timber frame are too heavy for one person to safely lift and carry around anyway. Instead, there are tricks to help roll, lift and move them using levers, fulcrums, pivots and log-rolling tools called cant hooks. Trestles and trollies are then used to swing and transport the wood. But having enough people around to help makes all the difference. The first step to saving your back is to stop and think about how you plan to move the heavy tie beam in front of you to where you want it. You then just get enough people involved in the lifting until it is comfortable to lift, and the job gets done together. Rachel Wilson, who attended a timber framing course, reflected that the most enjoyable part was going to the timber stacks in the mornings with the trolley and a group of people to collect the wood. It's a strangely rewarding experience. When six or more people are lifting something together, you can't really feel how heavy the wood is. You lift in time together and the timber feels ridiculously light. In this way the team also carried small tree trunks out of the wood by hand. I wouldn't recommend doing this on your own with even a small tree, as you quickly tire when the dragged wood snags on roots and dives into ditches. However, a group of people with a couple of webbing straps transforms the possibilities and makes a surprisingly powerful heavy-horse equivalent.

The skilful act of making joints in wood by hand is a more obvious ingredient in making an eighteenth-century barn. The hand tools – saws, axes, chisels – have a long tradition of use stretching back to the Middle Ages, the Romans, the Celts and beyond. The tools that we used for the Carley Barn were the same as those of the carpenters of Farmer and Sons in Royston and the original carpenters of the Great Barn at Wallington. Until recently, the knowledge of how to use the tools was passed from father to son, old-timer to young apprentice. It was a matter of demonstration, learning and practice. A big part of learning is tactile, using our hands, not just our eyes and ears. Only recently have we attempted to communicate this through books, videos and online. Information is now spread very widely, but this sort of mass broadcast has little sense of context. We no longer simply learn from our fathers and mothers, immersed in a craft since we could first stand up. We have to deliberately seek out knowledge and experiences.

Learning carpentry as a craft skill is a process of initially talking yourself through each part until the actions become second nature. This way of learning is accelerated, or perhaps intuitively made possible, by working with someone who already knows how to do it. I have always welcomed the opportunity to work alongside an experienced craftsperson, or to go on a course to learn a new skill. There is nothing more empowering than the absorption of a skill from one person to another. It feels like a very deep part of our human nature to pass on and receive manual skills and abilities. Our bodies internalise the actions until it no longer seems a conscious thing at all. Like riding a bike. We take this ability for granted, but it is a wonderful thing. Even the simplest things, like being able to run fast and stay upright. I remember as a child running through long grass and watching my legs moving quickly beneath me. I was awestruck by their speed, strength and ease without my having to think about my legs at all. I just kept running for the sight of the speed of their movement.

We think of these early memories of unselfconscious childlike responses as a state of grace. That state can still be felt in our bodies as we grow up. Later in my teens I began carving wooden plates, and after a while had a similar sensation of simply watching my hands working the gouge and creating the beautiful curved shapes of the plate as if they had minds of their own (both the plates and my hands). Letting go of conscious control, in a way being happily lost in your own body, is the kind of transcendent state that is often experienced by manual craftspeople. It is part of the joy of making. Our hands can produce perfect cuts with a handsaw to a fraction of a millimetre, and we can see the ¾ inch diameter of the head of a peg without needing to check it with a tape measure. This type of mastering of technique is a journey and takes practice, to be sure. But it is a natural thing for our bodies to do, and it seems to be achievable surprisingly quickly by some beginners attending timber framing courses. Over time, these abilities develop across a range of activities, until the process of setting out and cutting joints has a flow and virtuosity that feels very liberating.

The degree to which crafts are passed from person to person over generations is part of the 'way of life' of being a maker. Each person will build their own story of how they learnt their craft and who they met that showed them particular skills. Who was it that lit the spark of inspiration or opened doors to new opportunities? This is part of the web of craftsmanship. And those teachers will have their own stories as well, reaching back through time. That trail of connection is a lovely thing, and these stories are important when considering what the word 'craft' means today.

Sometimes there is a distinct starting point to a life as a craftsperson; sometimes it builds up over time. In writing this book I have realised that my own journey as a maker has been particularly influenced by two people. They were both very skilled but not necessarily skilled in the crafts I would eventually follow. One was the artist Alec Vickerman. Alec's ink and wash sketches were often made with a goose quill pen. He showed me how to cut the points, and his sons and I gathered the feathers from the marshes around their home. If we were lucky, we came across a swan's feather, which would be highly prized. I was struck by the free and slightly random expression of line that the quills gave to his landscape drawings, using such a crude and ordinary thing as a feather. Alec had a dry Yorkshire wit and his advice about life, and perhaps drawing as well, was to 'pay attention to the edges of things and the middles will look after themselves'. It applies to timber framing as well. Pay attention to the joints at each end of a piece of wood and don't worry about what happens in between.

The other person who greatly impressed me early on and inspired me into craft worked with clay. As a child my family briefly lived near the now famous craft potter, Richard Batterham. At our home, and when we visited Dinah Batterham's kitchen, we ate from Richard's plates and drank from his mugs. Richard and Dinah had worked at the Leach Pottery in St Ives, Cornwall, with Bernard Leach, and then built the pottery and kiln in a Dorset village that became their lifelong home; raising a family just a short walk from the potter's wheel. After leaving school, I stayed for a few weeks with Richard's family. Their thatched cottage was part of a smallholding with vegetable gardens, geese and bees. I didn't realise it at the time, but I was deeply influenced by Richard's philosophy of making pots as a way of life. I visited him again recently to better understand his philosophy and to be able to put it into words.

---

I met Richard in his kitchen, by the old Aga; I recognised the same stoneware pots, jars, plates, mugs and various teapots dotted around and stacked on the wooden dresser. All still clearly in everyday use. Very few people today can say they have spent their entire adult life working with their hands in direct contact with their chosen material. When he began working, making pots was a way of life you immersed yourself in. 'It's about the relationship with the material, the life story of the material,' he told me. 'Start at that beginning. Then, when you throw the clay, you don't know what is clay and what is hand. You are part of it. Not imposing yourself, your will on it, but part of it.' Very unusually, Richard doesn't make his mark on the pot's base. No initials. He would rather see 'a nice batch of mugs' rather than a pot that says 'me, me, me'. For him, the pot is what counts, not the potter.

Richard Batterham has made regular everyday platters in their hundreds. Each bowl was an attempt to achieve an almost intangible quality. It wasn't perfection, but something more important and more beautiful. Each pot he threw on his wheel was superficially the same as the previous one, but also unique in that moment of drawing the clay up. When the process is right, for the maker it would amount to being in a state of grace: a fluency and mastership beyond individual intervention. Something like this happens at moments during the scribing and cutting of carpentry joints by hand. You forget who you are and your hands and arms work with a knowledge and fluency all of their own.

Richard's goal was to form the shape, surface colour and texture that, after the haphazard effects of glazing and firing, would 'sing' to the owner of the pot. I see a similar thing in the components of a frame that has been well made. The happy accident of a run in the glaze is like the imperfect features in the materials in a barn. The skill of the carpenter is to not kill that beauty whilst at the same time using a focused capability so that the joints fit together like a hand in a glove. When the tenon slides effortlessly into the mortice and when the shoulder surfaces are exactly in contact, there is a satisfying 'klumpf' sound. The shoulders can fit so well that a cigarette paper cannot be drawn between them and the frame of rough-sawn curving timber seems to smile. This is part of the joy of the craft.

Of course, mistakes are made. But the barn aesthetic easily accommodates them, and it is healthy to not get too obsessed. At Barn Club we have a saying: 'It's a barn, not a clock.' Barns are unashamedly what they are. Nothing is covered up. Repairs that are made over the years are there for all to see. But the buildings remain beautiful, nevertheless. Carpentry mistakes in a barn are part of its character. They match the often less-than-perfect local materials, be that stone, earth or hedgerow wood. Mistakes are signs of human learning, and since we all make them, they are visible expressions of humanity. They make the record of a moment's hesitation or distraction. A misplaced peg hole, left empty with the corrected hole next to it, will be a feature that everyone is drawn to for decades to come. Rather like the fingerprint in the wet clay on the back of a handmade roof tile, which becomes the one that holds our attention the most. The mistakes let us know they were made by a human being. No one learning carpentry tries to make mistakes, but we learn to live with them, and in a barn they aren't so bad. For this reason, complete beginners and the carpenter-in-charge can both relax.

Having said that, every one of the fifty or so people involved in Barn Club were able to pick up the skills surprisingly quickly. I am continually impressed at the ease with which people can feel comfortable with the tools and make progress in spite of having no previous carpentry experience. It is a type of woodwork that is more accessible to most people than joinery or cabinet making, or any other form of woodwork that I can imagine. Rob Cullen, a landscape gardener with a gift for working with natural materials, expressed to me the delight of learning to use a drawknife to make the pegs out of freshly cleft oak. He said he couldn't tire of making pegs, once he got into the flow. Like Richard's pots, each peg is roughly identical but is individually made, requiring concentration and co-ordination between hand and eye to control the oak shaving as you pull the drawknife towards you. Rob was content to spend all day making pegs, with his dog dozing at his feet amongst the shavings. It is possible to buy pegs ready-made and delivered. But when it rains, retiring under cover to make pegs is part of the framing process. This way, you can also set aside the mistakes, the slight seconds and 'shorties', to use on the less structurally significant timbers such as studs. Very often at the end of the Barn Club days, people didn't want to stop what they were doing. Being satisfyingly tired is different from feeling exhausted. Some days you would stop and see that time has flown by and it's 7 pm already. This is a sign of healthy work and how engrossing the carpentry can become.

If we could bring together the components that distinguish this approach to carpentry, what would this sort of craft be called? It is a struggle to find a simple verb or noun in English. In the German language it is a little easier to express new concepts. Existing words can be strung together to form a credible new word with a new meaning. The German word could be something like *Volksnaturhandholzwerk*. This would be crudely translated back into English as 'People-nature-hand-wood-craft'. Unfortunately, in English this word is a mouthful! Its best left in German. But in the past the connotations of meaning were all there in what was experienced by a rural barn builder. The closest word or phrase I can find is simply 'village carpentry', borrowed directly from the title of Walter Rose's book, *The Village Carpenter*.

The answers to the questions that came to my mind in Wallington, when I stood and gazed up and around George Orwell's Great Barn, were revealed as the months went by. Our *Volksnaturhandholzwerk,* our 'village carpentry', was explored and developed through Barn Club. In it we found a wealth of experience that we were not expecting. This book attempts to represent the ingredients of that wealth, hoping to make it easier and possible for other people to try something similar.

Vince once related a story. A tourist in County Mayo in Ireland stopped a local man on his bicycle and asked him the way to Dublin. The man thought for a bit and then said, 'Well, if I were you, I wouldn't start from here!' So the trick in every journey is to know where to start from. Find that place and the rest falls into place. The place for a woodworker to start is trees. A love of trees. For the members of Barn Club, about to build an elm barn, this specifically meant elm trees. The next chapter is dedicated to the elm trees.

THREE

# Wych Elms, Field Elms and Wildings

The elm is an enigmatic tree. It is intriguing, often confusing and mysterious. Elm is a tree of paradoxes, a tree of surprises. It is also a tree that has suffered a pandemic fungal attack that killed large and small elm trees across the Western world. More than twenty million were killed by Dutch elm disease in Britain in the 1970s and the landscape hasn't been the same since. So perhaps the biggest surprise is the fact that there are now so many elm trees, even very big elm trees, around us at all. To discover a large elm tree near where you live is therefore like finding a missing masterpiece in an attic. You can't help feeling utterly astonished.

Barn Club volunteer Pete Cutforth described to me the pleasure he felt following his discovery of elms in a park in Edinburgh. Puzzling the identities of trees he was walking under, he found what he thought were lime trees:

*Lime trees, I thought, based on the bark, but the leaves out of reach, high up, looked more like elms. If they were elms, I couldn't believe my luck! Compared to the hedgerow-size elms I was used to seeing, these were proper trees, a rare sight for me. I walked on, wondering if I was mistaken? Further into the valley I spotted another group of potential elms. This time I took some photos, and could reach some leaves to make sure. Buoyed by the certainty of my discovery, I wanted to share it, shout out about these survivors, hiding in full view.*

Information about elms in books and on the web is extensive, but also often contradictory. Without firsthand experience to validate the claims, this makes writing reliably about elm trees particularly difficult. The approach I have taken is as much as possible to pare back elm facts from elm fictions through what I have encountered in my own lived experience. For example, elm wood has a reputation for being difficult to split with an axe. So much so that it is universally taken for granted. Also, the accepted carpentry tradition is that wych elm was preferred for making chair seats because it is strongly cross-grained and the chair leg sockets don't split. Yet, wych elm wood is also supposed to be the most easily cleft, or split. There appears to be a contradiction. The answer is to try it out, to try splitting elm and observe what happens. You will learn by sweat and effort to split your elm whilst it is still fresh and green, because as elm dries it gets considerably tougher and harder to cleave the wood fibres apart. In the light of additional context, both statements are correct.

Another example: there is an old English saying that beans should be planted when elm leaves are the size of squirrels' ears. It's a curious piece of wisdom and worth thinking through. As a historical saying, it undoubtedly refers to red squirrels rather than grey squirrels (grey squirrels being an invasive species that has now largely replaced red squirrels in the English landscape). Red squirrels' ears have several small tufts at their tips. And, sure enough, the emerging leaf bud of elm trees in spring presents not just one leaf but a cluster of leaf tips that will quickly grow on to become a sprig with leaves. For a period of just a few days, this emerging leaf cluster does indeed resemble the size and shape of a red squirrel's ear.

In Great Britain elm trees have always had a very close association with human societies, right back to the Bronze Age. Until a century ago, along with oak and ash, elm was inseparable from human culture, rural economies and physical day-to-day artifacts. Over hundreds of years landowners had preferentially managed the landscape so that these three species completely dominated the countryside. The highly productive hedged and wooded landscape must have been an amazing sight. English elm in particular spread through Southern England. Its enormous size, reaching 10 feet in diameter, meant it often rose above the surrounding trees by 20 feet or more. Its outline of a billowing cumulous cloud also made it easily identifiable. Most people who are over sixty today will remember giant elms on their village greens, bounding fields or towering in the churchyards. Long-lived residents will take you to an exact spot and point to the ground to show where the huge English elm trees used to grow.

I can remember working high on a roof in the shade of an English elm as a teenager. It was an imposing tree, with great descending branches, an ocean of small fluttering leaves and a vast, cool cavern of space between the spreading limbs of its crown. Its presence was impressive and its 'elminess' immediately experienced when working in its shadow. The tree had a tangible presence.

Since the 1970s there has been a catastrophic decline, not just in numbers of large elm trees, but a corresponding decline in our awareness and relationship with the tree. Along with this, elm's use as an everyday timber has dropped to almost zero.

However, times change, and our society is redefining itself, both in its physical sustainability and also its emotional connection to nature. Elms, as a part of our natural environment, are making a comeback. This revival, and our growing awareness of the importance of trees, is symbolic of a wider movement of appreciation and love of the world around us. A change of heart is taking place. Not just in awareness of the elm trees, but also in a renewed appreciation of the use of the wood, such as the timber frame made by Barn Club.

This revival was spearheaded a few years ago by the creation of an 'elms map' by the Conservation Foundation based in London and the UK website Resistant Elms established by the late David Herling. Private individuals on farms and estates, alongside the work of national botanical gardens, arboretums and certain charities, have preserved and planted collections of elms with apparent resistance to Dutch elm disease. There are field trials like those of Andrew Brookes for the UK Butterfly Conservation charity. Although there is no centralised elm initiative in the UK at the time of writing, woodland and forestry organisations are beginning to look at elm once more. The Future Trees Trust recently commissioned Karen Russell and Richard Buggs to review the 'state of play' of elms in the UK. Privately funded field trials of new fully Dutch elm disease-resistant elm types are also underway. The advent of genetic research now means that whole-genome sequencing for elms is theoretically possible. Elm-planting projects are not yet occurring at scale in the wider landscape, but the identification and testing of resistant elm stock by plant breeders, and their propagation at plant nurseries is surely only a matter of a few years away. There is still a lot to be done, but at least there is a sense of something possible to reverse the calamity that affected one of our most historically important trees.

All that said, the extraordinary thing is that the surviving elms are still out there. For me, the rediscovery of local elms and the pleasure of working with elm wood has been like stumbling across a secret walled garden, overgrown with brambles, and waiting to be discovered and brought back to abundance. For fifty years, elm has been a secretive tree, and its timber a hidden treasure. Some elms are so large and tall that when you see them it really challenges your rationality. They really shouldn't be there. They should be dead. They have completely survived the onslaught of Dutch elm disease as if it never happened at all. Through talking to other elm enthusiasts, I would put a figure of about one thousand elm trees in the UK with trunks over three feet in diameter. For trees of more than two feet, that figure can probably be multiplied by ten. There are even elms with a diameter larger than three feet, and elm pollards that survive as wide, squat, hollow drums that continue to sprout leaves. This feature occurs in other tree species such as oak, ash and yew trees which have had a fungal attack that rots out the inside of the tree, its dead heartwood, leaving just the living, growing sapwood intact. The hollow elm at Whaddon churchyard, Cambridgeshire, is one example. These rare, wide-diameter surviving trees don't match the tens of millions there used to be, but elms are definitely out there and very much alive. Many of these were clearly alive before the 1970s when the deadly Dutch elm disease swept through England. As botanist and 'elmer' Max Coleman puts it, if you cannot touch fingers when your arms encircle the tree, it is a survivor. And these are trees of great interest. So how have they survived?

This is a mystery still to be fully solved. A visit to the village of Boxworth in Cambridgeshire will reveal a small wood full of healthy, tall elm trees. It is an extraordinary sight if you aren't used to it. Why do some elm types persist as large trees? The observable range of appearance of elms that survive at least to semi-maturity is also baffling. The survivors aren't just one type of elm. I must admit that I like the fact that elms are so difficult to pin down. I like their spirit. It's like grain of the wood: very wild and unpredictable. In East Anglia there are other thriving elm woods like Boxworth, and countless spinneys, hedgerows and forgotten corners of fields with thriving semi-mature elm trees. All the elms that grow beyond about a 10-inch diameter are special. In the East Midlands and East Anglia, there are thousands of them, and each year the number is growing. Due to their relatively small size, they have probably appeared since the outbreak of Dutch elm disease, yet have lived beyond the age that elms are typically infected by the disease and die.

Any long-term resistance to the disease, or acquired field resistance, is not yet scientifically proven. But, nevertheless, there are increasing numbers of elms that are 10 to 14 inches in diameter. These trees would provide all the different timber sizes needed for a Hertfordshire barn, which is all that matters to a traditional timber frame carpenter. The significant observation is that, although larger trees are needed for sawlogs for milling into boards for flooring, chair seats and cabinet work, the smaller trees, unfit for conventional milling, will work perfectly well as structural framing timbers. Although there are relatively few large living trees suitable for cabinet work, there is a growing number of smaller ones suitable for timber framing. I should add that any very large living elm tree should be left alone and cherished. Fortunately foresters share this sentiment, so consequently there is no longer a ready supply of wide elm boards in England.

The region surrounding Hertfordshire has a diverse geology of chalk, clay and sand, and the gently undulating valleys of the rivers Great Ouse, Cam and Lea, and their tributaries, provide good growing conditions for a range of elm trees. The East of England has the greatest diversity of elm types in the UK and according to some sources, the entire planet. According to botanist Brian Eversham, who is developing a key for elm identification for the Wildlife Trust charity, the diversity of elm types in Hertfordshire runs to about fifty different types. I could believe it. My experience is that there can be variability in the appearance of elms and their timber from one hedge to the next and from one end of a village to the other. Fully grown elm trees differ in size, shape, limb structure, openness of their crowns, twig size and patterns, leaf shape, size and texture, and, of course, vulnerability to Dutch elm disease. There are different types that have been named, and, it seems, many more in between. It's not an easy business to look at a sapling elm and know for certain which type it will become and, if it survives Dutch elm disease, what form the tree might take in one hundred years' time.

The broad picture for the British Isles today is that there are two distinct species: wych elms and field elms, with numerous varieties and hybrids. Wych elms are truly native. That is to say, about 8,500 years ago, after the last ice age came to an end, they gradually spread across the landscape. Field elms are at the northern limits of their natural range and are usually considered native to the warmer climates of southern and central Europe. However, they are very much established in lowland Britain today and have been for over a thousand years. The two main UK species cross and produce hybrid types that may then cross again. Some are fertile and can spread both through their seed and root suckers. Others are not and spread through root suckers alone.

Early societies of human beings may have carried some field elm types here deliberately as they colonised the rivers, land and forests. Apart from its timber, elm leaves are a very useful livestock fodder, and it seems likely that some field elms arrived in the Bronze and Iron ages as people first discovered the British Isles. The tree we know of as English elm was introduced by the Romans and is now perhaps more correctly known by some as the Atinian elm, named after its original natural source near Rome. After the Romans, more field elm types were brought into East Anglia by the Anglo-Saxons. In the eighteenth century even more species of elm were brought here as specimens by explorers, plant enthusiasts and nurserymen. Some new elm hybrids were developed as well, such as the Huntington elm raised in Wood and Ingram's nursery in about 1750 from seed gathered from nearby Hinchingbrooke Park in Huntingdonshire.

As open communal pastures and heaths – the commons – were enclosed, and large private estates expanded, more nurseries developed for propagating particular elm types. These were then planted as a timber tree in new plantations in hedges and for decorative avenues. The eighteenth century is a fascinating period to investigate regarding the propagation of elm seed and root suckers. Authors such as William Boutcher, Batty Langley and William Ellis, wrote in detail about the prorogation, the planting out and the all-round benefits of elm trees. Following on from John Evelyn's book *Sylva,* published in 1664, they wrote about woodland management and documented the wide range of uses of the timber, from water pipes to chair seats. Elm was being promoted as a very profitable tree, more so than oak. There are accounts of extraordinarily fast growth and size. William Boutcher wrote in his 1776 book *A Treatise on Forest-trees*:

> *I have sold about sixty English Elms, at twenty-four years growth, of my own raising, at a guinea each, and not selected but taken in line: They were in general of about eighteen inches diameter, a foot above the ground, and forty feet high.*

A guinea is one pound and one shilling in old currency. A guinea in 1776 would be worth roughly £100 now, which means the value of each tree is about what it would be worth today, felled and stacked ready for collection. If true, twenty-four-year-old elm raised in the way that Boutcher suggests in his book would still be a worthwhile proposition today, when trees are normally assumed to need seventy years to reach a marketable size.

Propagation, planting and use of elm continued through the nineteenth century when the specimen trees in botanical gardens and stately avenues grew to maturity. Although during the same century our use of native elm wood declined relative to imported elm timber, particularly elm from North America. In the mid-twentieth century the distribution of notable elm trees and their enduring relationship with humanity was explored by R.H. Richens in extraordinary detail in his book, which is simply titled *Elm*. It is a tragedy that, just as his book was being published, an epidemic of a new, deadly strain of Dutch elm disease appeared on British shores. But this should not be seen as the end of their story. There are still many fully-grown elm trees hidden across the countryside and in our towns.

---

Elm trees are fun to try and find. The surest place to start, without doubt, is to carefully look at the trees you see every day. Tree guides will help, but can be confusing when trying to identify the actual species or variety of elm. The first thing is to be able to identify that the tree standing in front of you is definitely an elm tree. The shape of its leaf gives it away. They are an even oval, running to a point and the edges of the leaves are serrated. The most important feature is that they are asymmetrical at the leaf base. That means that the blade of the leaf appears at different heights on either side of the stalk. To identify the tree in front of you as an elm is therefore easy. To identify exactly which type is trickier. Elm leaves vary in size, overall shape, hairiness, shine and proportion. This diversity even occurs on the same tree. The best leaves to gather need to be fully developed from short stalks and in full sun, ideally from mature trees. Leaves from lower down the tree, or from young saplings are often hairy on both sides, even for so-called 'smooth-leaved' elm trees!

There is nothing more satisfying than to turn a corner in a country lane to discover a majestic elm, 4 feet across at its base, standing 70 feet high, with glorious sweeping branches and filigree leaves and twigs. Some of these trees are in their prime and are magnificent. Bearing in mind that most people think that the elms are now all gone, it is deeply satisfying to find a large or veteran tree breaking all the rules.

Spotting elm trees is a skill to learn, but very rewarding as they begin to take shape in the countryside around you. The signs to look for depend on the time of year and the distance you are from the tree. Ultimately, leaf shape in summer is the

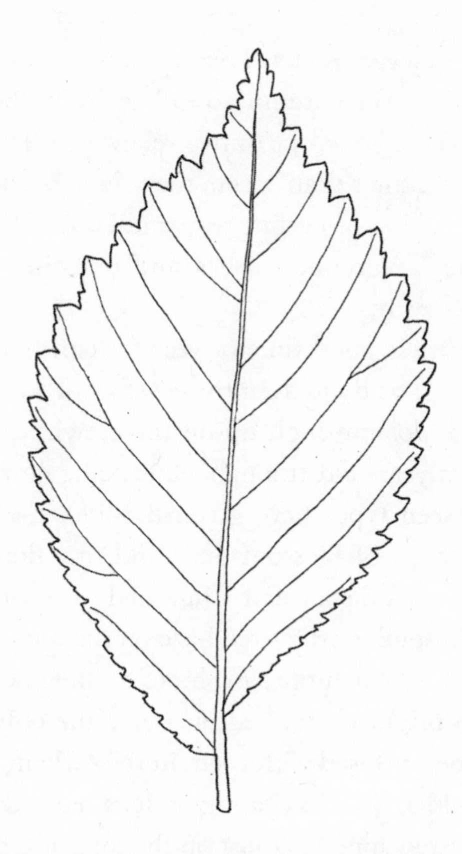

The distinctive a-symmetrical shape of an elm leaf.

deciding feature that will identify the tree as an elm. But, for this degree of observation you need to be close enough to pick a leaf. To find the elms at greater distance around you, you need to look for other clues. Strange as it seems, it isn't easy to spot elm trees in the landscape when the leaves are out. To begin with, in summer the density of leaves on trees and saplings in the foreground block out the trees behind. Elms also blend in with other tree foliage at eye level and are easily overlooked. There are so many elm leaf shapes and sizes, that they look similar to cherry, hornbeam, hazel, blackthorn or even goat willow. However, elms hold their leaves long into the autumn, so in late November, when all other leaves have gone, one or two yellow elm leaves hang on. Being yellow rather than green, they shine out like torches. If you are walking or cycling, you will be able to spot them through the foreground bushes and branches, even from several yards away.

Another really good time of year for elm hunting is late winter and can be done from the comfort of a car. Elm trees produce red blossom high up on the crowns of trees from about fifteen years old (though, elms being elms, some less commonly seen types have green flowers). Each tree produces thousands of blossom buds and tiny floret bunches. Observed from a distance of a hundred yards or more, they form a dusky-pink or ruby-red haze on the tops of trees that are easily visible when the sun shines on them. Occasionally I have seen bright cherry-red blossom. The only other tree that could be confused with them in the early months of the year is the alder, with its coating of deep red catkins, so look for the ruby-red appearing just on the tops of the crown. All the other trees around will still be dormant, so their branches will be bare. Elms blossom very early in the year from mid-January to early March, depending on the elm type and local climate. The earliest I have seen elm blossom on English trees is in late December. Frustratingly, it is difficult to see the blossom close up because it occurs on the tops of trees. But reaching up for a drooping branch in blossom to see the buds and flowers close up is well worth the trouble. The colour and complexity of form is beautiful. When cut and brought indoors, the branches will also leave a covering of pale primrose-white pollen on the shelf, minutely fine and light as talcum powder.

The other advantage of identifying trees in winter is that the full structure of stems, that is to say the main trunk and its branches, along with branchlets and corresponding twigs, become visible. However, it is easy to be mistaken. Elm trees have such a range of shapes and twigs that in winter, at a distance, it is easy to confuse elms with lime, oak, willow, hornbeam, beech and even field maple. To help, some of the field elms have what looks like a fluffy top as their twigs are very delicate and lace-like. Other types have more stubby and distinct twigs. This varies region to region. Once you have positively identified some elms in your area, you have your eye trained to spot other similar trees from their winter coat alone.

Perhaps the easiest identifier of all is when the elm produces seeds. After flowering and before bursting into leaf, elms produce thousands of bunches of seeds, looking like handfuls of confetti. Each seed has a flat, light-green round or oval wing surrounding a darker seed in the middle. In fact, dried elm seeds make the best confetti as they seem to hang in the air as they spin and fall. Again, this all happens before other trees produce leaves, and the crown of the elms briefly turns a distinctive milky, light green before the seeds are lost behind emerging green leaves. Even at a distance of a mile, the pale colour of the seeds is unmistakable. Once you have seen it you won't forget it. The seeds are also delicious as a salad, tasting slightly nutty and full of protein.

I recently discovered two giant smooth-leaved field elms, one with a trunk larger than 5 feet across, in the valley in Devon where I had lived for twenty-five years. Some of the combes and gullies at the edge of Dartmoor are isolated and sometimes even unnamed. They are part of the scenery, but the lanes and footpaths often bypass them. When you take a detour to explore an overgrown gulley, you feel that only wandering sheep and our ancient ancestors have ever set foot there. The rocks have become pillows of moss, and lichens hang down from the branches of the trees like ticker tape. The elms I found were just two fields away from a well-trodden path. Although about 60 feet in height, from a distance they looked like giant bottle brushes, more typical of the shape of Dartmoor beech trees. The elms were certainly wind swept and had multiple branches sprouting from short main trunks, dividing again and again into myriad small twigs. They were only eventually distinguishable from beech trees after a closer inspection.

I was amazed that I had lived in the same valley as these huge elms for so long, looking at them almost every day from the valley bottom without recognising them. It was a real thrill to discover them. As far as I am aware, these elms were only known to the farmer who owns the land. Some trees like these might have an exact botanical identity and even a common name, but others may not yet fit a textbook description. They may be geographically isolated and genetically unique, their vernacular name fallen out of use, even if there was one in the first place. They stand as part of the landscape that we think we know, only recognised as elms by perhaps one or two living people.

One February day on the other side of the country, I came across four large elms lining the bend on a road in the village of Keyston, Cambridgeshire. From a distance they looked like oak trees. So I did a double-take as I realised they were definitely elms. They were all in good shape in 2019 and gave a good impression of what rural England used to look like, with lanes that were lined with tall elms with long drooping branches laden with leaves. Having found a large elm tree, it is worth exploring the immediate lanes and byways. If there is one good tree, there are probably more somewhere in the same parish, sometimes a similar large elm growing in someone's garden – lucky people!

Small, compact groups of the same type of survivor elm are the usual pattern in the landscapes of the East Midlands. An example is next to a layby near the village Melchbourne, Bedfordshire. It is a small copse showing no sign of disease, occupying 600 square yards in a corner of a field. Unfortunately, the site is currently used for fly-tipping rubbish. The site is not even marked as a wood on any map. There are about 250 trees in total, densely growing in this small area, which includes a large oak and some ash trees. It looks like the elm group has spread from one large parent tree, 3 feet in diameter and about 70 feet tall. This grandmother tree grows by a ditch, is covered in ivy, and has great spreading branches. Nearby are twenty-one mature trees, perhaps ninety years old, with deeply furrowed bark and lots of side-whiskers or burrs. These reach 50 to 60 feet tall, and range between 14 and 24 inches in diameter. Some of the elms had multiple stems, or trunks, growing from the same stool, or shared base, suggesting they were cut back or coppiced when young. The density of trees has forced them to grow stems free of lower side branches and each tree is topped with a small crown of branches, meaning the growth rings of the timber inside will also be tight together. There are a further 20 young trees, 4 to 14 inches in diameter, and at least 200 root sucker saplings in between.

There are also some cities rich with mature elm trees, such as Brighton, Edinburgh and Amsterdam. In many other towns and cities, perhaps most cities, there are occasional individual large elm trees as well, such as in Sheffield and Stevenage. When you are going about your business, they appear out of nowhere. So if you are an elm hunter you need to be permanently on your guard.

---

Finding an interesting elm is one thing. Knowing exactly what species or type is quite another. A comparison with the taxonomy of apple trees in the UK is helpful. Through DNA testing there are known to be about three thousand different named varieties of apple trees growing in the UK. There are perhaps a further two thousand yet to be fully identified and named. The reason is that every apple that contains fertilised pips will produce genetically different offspring to the apple tree it came from. Each pip that grows may have very similar characteristics to its parent, or it might be wildly different, producing a new apple with a unique appearance and flavour. For example, there is a popular cooking apple variety known as Bramley. Each Bramley apple tree is a clone grown from a cutting taken from another Bramley tree, which was also once a cutting, and so on, back to the original Bramley tree that was grown from a pip sowed by a small girl in 1809 in her garden in Southwell, Nottinghamshire. All Bramley trees all over the world are clones of that one tree. So, an apple core thrown out of a car window by a child in 2020 might germinate in the verge to produce a new variety of apple that tastes different, looks different and is indeed genetically different from its parent. Collectively these apple trees are known as 'wildings'.

Elms differ from apple trees in many different ways, of course. Their seeds have circular wings and are dispersed on the wind. But the big difference is that field elms don't easily reproduce by seed in most of the UK. They grow new shoots and saplings from where their roots are spreading, like an underground version of a strawberry plant. I have found young root suckers growing in a meadow at 40 yards distance from the parent tree, though they will normally sucker within just a few yards.

According to Richens, field elms successfully cross-pollinate and germinate in the wild in Great Britain only about once every fifty years or so. Although accounts of elm propagation in nursery conditions from the eighteenth century don't mention any difficulty, nor in 1913 did Henry John Elwes and Augustine Henry, writing in the seventh and last volume of their comprehensive study of trees, *The Trees of Great Britain and Ireland*:

> *If it is desired to produce new varieties, the raising of elm from seed is a very simple matter, provided that the seeds are sown as soon as they are ripe, when they germinate in a few days, and make strong plants in the first year. The variation of most elms that I have sown, except the wych elm, is very great, and natural cross-fertilisation no doubt accounts for this.*

Of course, climatic conditions change over time and future changes through global warming might provide conditions better suited for elm pollination and seed germination. Once a field elm seed has germinated and grown into a young tree, the tree can spread through root sucker shoots. As clones they share the same DNA as the parent tree and so are only able to display features unique to that elm tree type. The clones might slowly spread along a hedge or the edge of a wood, but unless their root suckers are deliberately dug up and transported by a helpful human, they will only spread very gradually from the place of their origin. Wych elms, on the other hand, only spread through their seeds and they do not root-sucker. Thought of in this way, it is possible for there to be hundreds of different elm wildings along the lanes, roads and hedges of lowland England.

Typically, elm trees like good, deep, well-drained soil. But there are also elm types that survive on heavy clays or shallow chalky soils. They seem to like the disturbed ground around ditches, embankments, hedges, tracks and lanes, all common in the farmed landscapes of Hertfordshire. Elms also spread into the edges of woods, often the southern edges where it is warmest, though they can appear almost anywhere.

Propagation of field elms from a large parent tree by root sucker shoots is surprisingly straightforward. It is done in late autumn or early winter, ideally in November when the tree is at its most dormant, which allows the roots and earth to settle before the first leaves flush in the spring. A 2-foot-tall elm sapling only needs a few inches of a parent root the width of a finger to survive on replanting in a nursery bed. With good topsoil and regular summer watering, they will grow to 6 feet tall in 2 years and are then ready to plant. Sapling field elms can thus be transported great distances, which makes it easy to imagine that some of the field elms came to the British Isles through human agency. In the eighteenth century elms even as large as 40 feet tall were transplanted.

Elms can grow well in much tighter densities than other trees like oak, and the volume of timber per square yard of woodland is consequently greater. In the eighteenth century this was understood, and also that elms thrived best in relatively narrow strips of woodland, at the time called 'hedge rows' where their roots could spread out into the fields beyond. There are many examples of these in the East of England, and often a strip of land just 10 yards wide, and unmarked on Ordnance Survey or Google maps, will host 4 or 5 rows of healthy elm trees.

The copse at Melchbourne is a typical example of a small elm group. Although enough light is getting to the trees from above and on all sides to nourish them, the copse is unmanaged. So felling a few trees would help the remainder grow to their full size. From a carpenter's point of view, those thinnings would generate very good timber-framing material at no cost to the health or continuation of the copse itself. As is often the case, the owner did not realise that they were elm trees. He believed the elms to be burr oaks. This copse is also a good example of how elms persist unknown and largely unrecognised. It is likewise safe to assume that there could be more elms growing closer to your home than you might think.

Interestingly, in the past it was believed that, unlike oak or ash, large elm trees growing in hedges would not be to the detriment of the surrounding ground's productivity, which could continue to support normal growth of grass – a proposition repeatedly claimed by different eighteenth-century authors and it would be fascinating to establish the truth of this claim.

---

It is understandable that most people who own the trees do not know the name of the individual elm type. According to which tree guide you look at and when it was written, the names of the elms might vary, which adds some fun and confusion. For me the most important thing is that the tree you are looking at is an elm, a healthy elm and one that looks like it is going to become a large tree. I recognise distinctly different types when they are mature trees and name them in my own mind after the farm, village or region where they grow, such as the Whaddon elm or the Melchbourne elm. I imagine that people in the past did the same thing.

I personally admire the fact that elms are living things that defy an easy understanding. Elm taxonomy, the botanical classification of elm trees, is complicated and generally accepted scientific names, as well as common names, have changed over the decades. It is one of the most difficult plants to identify. The full list of elm types in Britain currently extends to some sixty-two types of elm, as identified by Peter Sell and Gina Murrell in volume 1 of *Flora of Great Britain and Ireland.* There are two schools of thought amongst botanists as to whether these are multiple distinct species of elm (this school being the 'splitters') or multiple varieties and hybrid crosses of wych elm and field elm (the 'lumpers'). When whole-genome sequencing of all the elm trees has eventually been conducted, the matter may be resolved. For this book, I assume there are two main species, with different varieties and also hybrid crosses between them.

So, in more detail, firstly we have the wych elms, *Ulmus glabra.* Wych elms thrive where the summers are cooler, though they seem to be present in some shape or size in every county. At first sight their leaves look just like hazel leaves, with their round to oval shape, pointed tip and double serrated edge. But as soon as you touch them, you feel the very coarse hairy surface. In contrast, hazel leaves are soft and smooth. The north and west of the UK are traditionally the havens for wych elms, where it was the dominant species of elm and there is more chance of seeing a mature wych elm tree in these parts of the countryside. In Scotland there are still many large wych elm trees and some of them are tall, valuable timber trees, perhaps including those referred to in the past as mountain elms. At Craiglockhart in Edinburgh, there is a 100-foot-tall wych elm with a good straight-ish trunk about 3 feet in diameter. Down in the south, wych elm doesn't any longer grow anything like as big and tall. The wych elms growing in Hertfordshire grow no bigger than shrubs before they are knocked back to their roots by Dutch elm disease. There are lots of them in the hedges, often in close proximity to field elms such as juvenile English elm or other hybrid elms. The larger wych elms that do survive here often grow into tall, spindly trees hidden amongst other tree species in a wood.

Wych elm, our native elm, has a name that is ancient. 'Wych' is a seventeenth-century word, from the Old English *wice,* meaning 'supple' or 'pliant'. In the context of the Anglo-Saxon world, the name 'wice' might relate to the qualities of the wych elm tree as it was used. As mentioned at the start of this book, the wood itself can be cleft when green and is particularly pliant in thin sections. Boards and battens of wych elm can be bent into shape more easily than any other wood, including ash. Likewise, the layer of fibres immediately under the bark, the bast, is strong and flexible, and can be easily made into string and rope. It can also be cut into strips and woven into chair seats, or a basket. This is a rare property only shared with the bast of lime trees. Of course, today we don't often use woven bast baskets or natural fibre string and rope, so we have lost sight of the significance. We have fibreglass boats, plastic buckets and polypropylene rope. But viewed from the perspective of only using local natural materials, a tree that has very useful bendy properties might be named accordingly.

The other species of elm in Britain is the field elm, *Ulmus minor.* A large number of field elm trees in England today are normally seen lined along roadsides as groups of dead-standing young trees, stripped of leaves, twigs and bark, just grey poles and very much deceased. Repeated observation of trees like these and their stark image of desolation, along with the historic use of elm to make coffins, feeds into the general impression of the sorry state of elm and emphasises an association with mortality. Yet, given time, and the selection and planting of disease-resistant varieties, we can look forward to a future that is very different. Rather than always thinking of elms in the past sense, in the way of an epitaph, we might think of them as a species that has suffered a pandemic that can be recovered from. Our knowledge of what field elms were like before the disease took hold, can lift our vision of what might be possible again. So, with this in mind and resisting nostalgia, we can explore what those elms meant for our country in the past and now have within our gift to restore.

Field elms were a quintessentially British tree, experienced every day in our winding lanes, village greens and urban parks. They framed our farmed landscapes, and were the tree of choice for our greatest stately avenues. We should also remember the hollow stems and branches that were homes to owls, jackdaws, woodpeckers and a whole host of other creatures, including bees. Even the leaves were important for our ecosystem. When trying to find a good specimen to photograph, I found most leaves seemed to have had a bite or two taken out by some sort of resident caterpillar. Mature elm trees support a very large number of living species, from fungi to butterflies, insects and birds. Oak is famous for this important quality, but elm is not far behind, supporting far more forms of life than ash trees, beech, hornbeam or sycamore. The white-letter hairstreak butterfly is dependent on elm, its caterpillars hatching from eggs as the blossom arrives, beautifully disguised in shape and colour, changing with each larval stage to mimic the buds, flowers, seeds and young elm leaves as they appear in sequence.

If a large elm tree loses a branch in a storm, the wound is exposed to numerous types of air-blown fungal spores that establish and grow, feeding on the timber and rotting it away until a hollow is formed. But elms have a remarkable ability to grow sapwood to bridge back over and, over a number of years, seal the hollow. Once the cavity is sealed, the rotting seems to be arrested and the tree continues to grow as normal. In the meantime there is a nice, dry space for birds to nest, entering the cavity through the dwindling access hole. Likewise for insects. Wild honeybees build colonies inside hollow elm trunks and branches, and the UK may well have lost tens of thousands of natural honeybee hives during the devastations of Dutch elm disease. I still encounter wild honeybee colonies in surviving elm trees, one of which had been occupied by bees for twenty years before the tree's sapwood grew back across the access hole. Another fallen branch from an elm in a Cambridgeshire village exposed a honeybee colony within that hollow stem. The owner did not realise, which is very normal. The bees were too high up to be seen or noticed. As with the potential existence of large elm trees nearby, most people are unaware of wild honey bees living high in the trees around them.

The early budding of elm tree blossom can also provide food for birds. This winter I counted twelve goldfinches grazing on the blossom buds in the crown of an elm tree. Against the brilliant blue of a clear sky, it was a beautiful sight. These trees really are gentle giants, providing homes and habitat for the species that we find we now care most about. When I see a large elm I always try to speak to a local person who knows anything about the tree. Very often there will be a reference to wildlife. One family in Essex has a beautiful pair of large elms at the end of their drive. They have long sweeping branches, and the owner calls them the sentinel elms. 'Oh,' he said as I was leaving, 'and there are little owls living inside, up there in that hollow branch.'

The English elm, the type of field elm that dominated the English landscape, was used for its timber, as you might expect, but its leaves were also valuable as fodder for livestock. The use of hedge trees for fodder is still locally known as 'pannage' in Devon. The tree stems were also used for growing vines by the Romans; this combination of tree and vine was intriguingly referred to as 'marriage'. The vines were believed to grow better and the trees continued to produce fodder, so it was a perfect combination. English elm trees often occur in paintings of idyllic landscapes of rural England, such as those by Constable, and are associated with romance in poetry as well. If the historic cultural associations of elm can be seen to link it with beauty, pastoral scenes, love and even marriage, then the elm has positive connotations that we can also relate to.

---

Elms themselves also have a tremendous will to live. When we stacked small-diameter elm tree thinnings ready for milling, some stems sprouted new shoots and leaves from the smooth bark even nine months after felling and in the middle of a dry summer. The capability to vigorously grow from root suckers also means that one lone elm tree can spread to become a dense clump. But the clonal nature of that reproduction means that the DNA is shared, and this is the tree's Achilles heel. If a new pathogen arrives and breaks the tree's defences, all trees with the same DNA, such as

English elm, will equally suffer. When a new, alien fungus appeared in the 1910s and again in a more virulent form in the late 1960s, countless cloned elms became vulnerable.

The fungus is now suspected of being endemic to Japan, brought to the UK indirectly via Canada on the bark of imported rock elm logs destined for dockyard piling. In the past rock elm was imported in squared, hewn baulks of timber, which guaranteed the bark, and therefore the fungus, would be left behind in the port of origin. When bark was left on the imported logs, an alien fungus, *Ophiostoma novo-ulmi,* also arrived. Most of our elm trees, particularly English elm, had no means of resisting this new fungus. Normally, in mature ecosystems, trees, fungi and bark-boring insects coexist. Where they have co-evolved, none of them dominate at the expense of the complete destruction of the others. However, when alien trees, fungi or insects are introduced from distant continents, that essential process of gradual co-evolution is not present. There is the potential for killer fungi, killer insects or an uncontrollable spread of invasive plants. *Ophiostoma novo-ulmi* is one of these. It is a variation of a native fungus, one which is far less damaging to our native elms.

The fungus is carried very effectively from tree to tree by two types of scolytus beetle, our native elm-bark-boring beetles. The adult beetles feed on sap as they burst into leaf. In early summer the beetles bite through the thin bark at the 'V' joints between the twigs. It is then that the *Ophiostoma novo-ulmi* spores on the mouthparts of the beetle pass beneath the natural defences of the elm tree and spread through the xylem vessels (the tubes in the sapwood that transport water and minerals) to infect the rest of the tree. The elm tree responds by exuding a brown sticky chemical into the sapwood vessels, which also has the effect of clogging them, and without water the leaves turn yellow and die. When the whole tree has died, either in the same year or the next, the beetles return to lay their eggs beneath the bark. The larvae hatch and eat their way along the cambium layer (between the inner bark and the sapwood) to create radiating galleries of small tunnels. These fanning galleries are extensive and easily visible if you prize off the bark of an infected dead elm tree.

In the early 1970s the impact of the newly arrived fungus was not fully appreciated. It was called 'Dutch elm disease' as it seemed to be the same disease – caused by the fungus *Ophiostoma ulmi* – that had been identified earlier in the twentieth century. Ironically, it neither originated in the Netherlands, nor was its effect limited to the particular type of elm currently called Dutch elm. However, it was first identified in the 1920s by the Dutch scientists Christine Buisman and Bea Schwarz. The impact of this disease on our elm trees and landscape was like a tornado, but it is misleading and unfair to blame it on the Dutch!

However, as we have found, not all the elms were killed. The hope is that there is a large native elm tree somewhere that has a natural and complete resistance to the Dutch elm disease fungus. This is unlikely, but as elm researcher Karen Russell says, 'That needle is in an ever-diminishing hay stack!' There are many factors that give some trees degrees of tolerance to Dutch elm disease. Much more research is needed to explore the mechanisms of both resistance and tolerance within our native elms. Hopefully, at some point in the future, a parasite of the bark beetle or a virus in the fungus might incapacitate Dutch elm disease, allowing these saplings to grow to become great trees once more. But for the moment, depending on the type of elm, the majority of suckering elms reach an age of five to ten years, or grow to a height of between fifteen and twenty feet, and then are infected with *Ophiostoma novo-ulmi* and die.

The behaviour of the scolytus beetles as vectors are possibly the key to the question of why some elms avoid infection. The impact of summertime temperatures on the flight days of the emerging adult beetles is one consideration. The beetles wait for temperatures to be above 21 degrees Celsius to fly. Also, the time in the year that the elm trees start flushing varies. This is when sap vigorously flows to the growing leaves and stems, attracting the first beetles to feed. Slower-growing trees may also be less attractive for this reason. I know of two resistant elm types that begin to flush in late February and start producing leaves in March, long before the other elms. There are also chemicals given off by the leaves of some elm types that may be distasteful to the beetles, and other possible environmental factors, such as onshore salty winds or the height at which each of the scolytus beetle types naturally want to fly. Since the fungus needs the beetle to transport it to new host trees, keeping the beetles at a distance may be how some elms tolerate Dutch elm disease that is endemic within their local environment. I suspect the answers are complex and that there are multiple factors at play. Whatever the case, to discourage the activity of the beetles further, the bark of any dead and fallen elms should be removed and burnt because the beetles breed under the bark of dead elm trees.

Brighton is famous for protecting its remaining elms and hosts the National Collection. However, to see fully formed English elms, you might now need to travel to Melbourne, Australia, to the Fitzroy Gardens there. Victorian colonials must have planted some specimens from the 'old country', and they look just like the trees we used to have here – except for the enormous flocks of giant fruit bats that roost in their branches. Australia now has rigid biosecurity, and for the sake of those trees I hope it continues. A similar situation exists in New Zealand, where Dutch elm disease has not yet landed. Apart from old photographs of elm trees, you can also sometimes see large elms in the background landscapes of films set in the English countryside during the 1960s and 1970s; *Chitty Chitty Bang Bang* and *The Railway Children* are two examples.

It is a strange twist to the story, but Dutch elm disease doesn't necessarily kill each tree completely. For every large tree that dies, several new root suckers grow from its surviving root system. Alongside the familiar sight of dead-standing trees spread across almost every parish in lowland Britain, is the sight of fresh young elm trees, often nearby. It seems counterintuitive, but the number of elms in the wild is actually growing. Due to these multiplications there are now more elms than ever before, although I would guess that 99 per cent of them are saplings. A really good example of the explosion in numbers of sapling elms is along the embankments of motorways. The density of stems is astonishing, as tight as 4 inches apart. For example, these tightly emerging elm spinneys grow along the embankments of the M4 from London, and the southern sections of the M5 in Somerset are also home to thousands of young elms. It appears as if a recovery is underway. Only time will tell if any of these have resistance to Dutch elm disease. Where it is possible to stop in safety and inspect the trees at a service station, there are occasional elms in excess of 18 inches in diameter. Could any of these be new hybrid trees, forerunners of new disease-tolerant strains?

There is potential for an elm revival in the UK to be part of the 'rewilding' movement and the restoration of natural landscapes more generally. Elm trees were so much a part of a healthy, abundant natural environment throughout our history, that they should not be overlooked in our plans for the future. In the context of our society's struggle against environmental loss and climate chaos, the symbolism of the elm is something grounding and earthy. It is one full of complexity and contradictions, but also a symbol of renewal, of survival against the odds and of life after tragedy. A tree of hope.

Only time will tell, but the large numbers of usefully sized elms that have appeared and grown since the arrival of *Ophiostoma novo-ulmi* suggest something has occurred in the wild in the last fifty years. Even if the answer is 'no, the needle is still lost in the haystack', there is potential for field trials of new, fully disease-resistant hybrids. If these elms also produce well-shaped healthy trees, able to withstand storms and drought, there is hope that by 2050 an elm revival will be well underway. As well as a long-awaited return of majestic elm trees in the British landscape, a rediscovery of elm wood as a beautiful material for making timber frames, furniture and other useful objects will also follow. This re-emergence, this renewal as the elms return, is part of the reason for seeing elm as a symbol of returning life.

That the eighteenth-century barn in Wallington should be made of elm makes perfect sense in the historical context. For the new handmade barn at Churchfield Farm, the choice of elm as the timber frame made even better sense. Here was a tree that had been overlooked and forgotten, just like the old ways of working by hand and the carpentry techniques that we planned to use for the Carley Barn. But before we could set out into the wood to gather the elm, we needed to have a design in mind and know what the barn would look like. For the timber frame we had to know exactly what size, what shape and length of timber, and how many trees we would need. Just like the village carpenters generations ago, there was no point in finding and felling a tree until we knew what it was going to be used for.

FOUR

# Plinths, Plans and Preparations

The proof of the pudding is in the eating, as the saying goes. Likewise, the proof of a timber frame is in the raising. That is when you really discover if the hundreds of joints will fit together. The key to both a good pudding and a good timber frame is in the preparation. And there is a lot of preparation behind a well-made frame.

The entire structure needs to be drawn in enough detail to use as working drawings when the carpentry begins. These are the framing drawings. Then, there is the cutting list of all the timbers needed. When the logs arrive on-site they need to be calibrated and recorded on a log list. And, when the milling starts, the cutting list and the log list are brought together as the milling list.

Regarding the framing drawings, it's worth making an appraisal of the local traditions where you live. For the Carley Barn, this meant some detective work, investigating local historic frames for the detailed design of the barn. There aren't any copybooks of framing drawings that cover all the possible barn designs. So the best place to start is to explore, find and measure what barns there are nearby. Most farmers are proud of their old barns and will show you round or let you have a look and take some notes of materials used and any details you can see. Better still are folk museums and properties owned by charities such as the National Trust in the UK. Researching the agricultural vernacular is also a good excuse for a great day out.

I will never forget my first visit to the Weald and Downland Living Museum in Sussex as a teenager. The museum has a variety of original historic timber frames that had been rescued from demolition, restored and rebuilt on the gently sloping meadows of the valley in the South Downs of Sussex. Two things are stuck in my memory of that day. One was walking inside the Hambrook barn and the wonderful mixture of smells of old oak, thatch and limed brickwork. This was the first of several buildings, each one compounding the sense of wonder in me that so much could be done with such crude, raw materials that could be counted on the fingers of one hand. It was such a rich, earthy world, full of the fruits of local materials, hand tools and craftwork. Before I knew it, the day was over. The other lasting memory was the apple pie with a woven pastry top that my friend's mum had brought along for us to eat. It was an inspiring day and one that set the wheels of my career in motion.

The Chiltern Open Air Museum is local to where I live now. It houses a number of agricultural buildings available for close inspection. There are aisled barns like the Great Barn at Wallington, slightly smaller single-span barns, wagon sheds and stables. In the villages near my home there are also timber-framed cart sheds round at the back of some more ancient pubs. If you want to build a timber frame but don't know where to start, barns are great to investigate because their timbers are all exposed. These beautiful buildings on farms

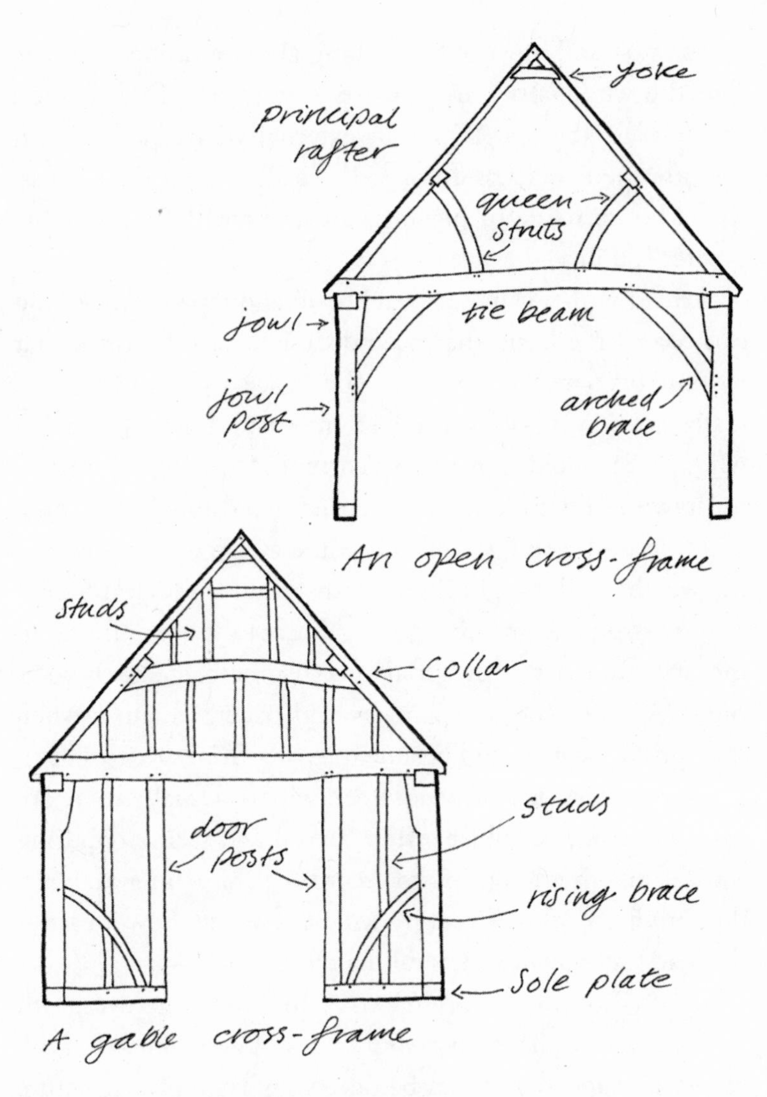


The components of the cross-frames.

and in folk museums are, in effect, instruction manuals. After viewing several barns, it is possible to begin to see repeated patterns in the materials used, the style of framing and the dimension of the wood used for different components.

The first and most obvious thing about a timber frame is that the whole structure consists of a series of flat planes: the frames. There are the four external walls, the roof and the intermediate cross-frames. These all meet at their edges, like a house made of playing cards, but with considerably greater stability!

Armed with a camera, sketchbook and tape measure, you can record the basic shapes, and dimensions. Features that reoccur become recognisable. The process of drawing is a really good way of focusing attention and actually seeing what you are looking at but not currently noticing. So it helps to draw each flat plane of the structure. Through sketching each piece of wood, you also become aware of the way that the weight of the roof is being transferred through different rafters, struts, beams and posts. Noticeably, every frame in the structure has a diagonal brace somewhere to hold it rigid. Without them, racking would occur, which is when the upright posts would be able to swing from side to side in the wind like a pair of windscreen wipers. Usually there are two braces in each plane, often curved and facing opposing ways. Through attempting to record the frame in a drawing, the simplicity of the structure will be revealed; how it carries the load and how it resists falling over.

The quality of wood often looks poor to modern eyes, with saw marks, axe marks and sometimes large splints torn off. However, appearances can be deceiving. In an old structure the species of tree was carefully selected to be right for the job. To eyes used to clean square pieces of wood, the number of knots and amount of bark still on the wood is remarkable. But don't think that the craftsmen didn't care. The structures wouldn't still be here if the carpenters didn't know what they were doing. They knew what would work, where the

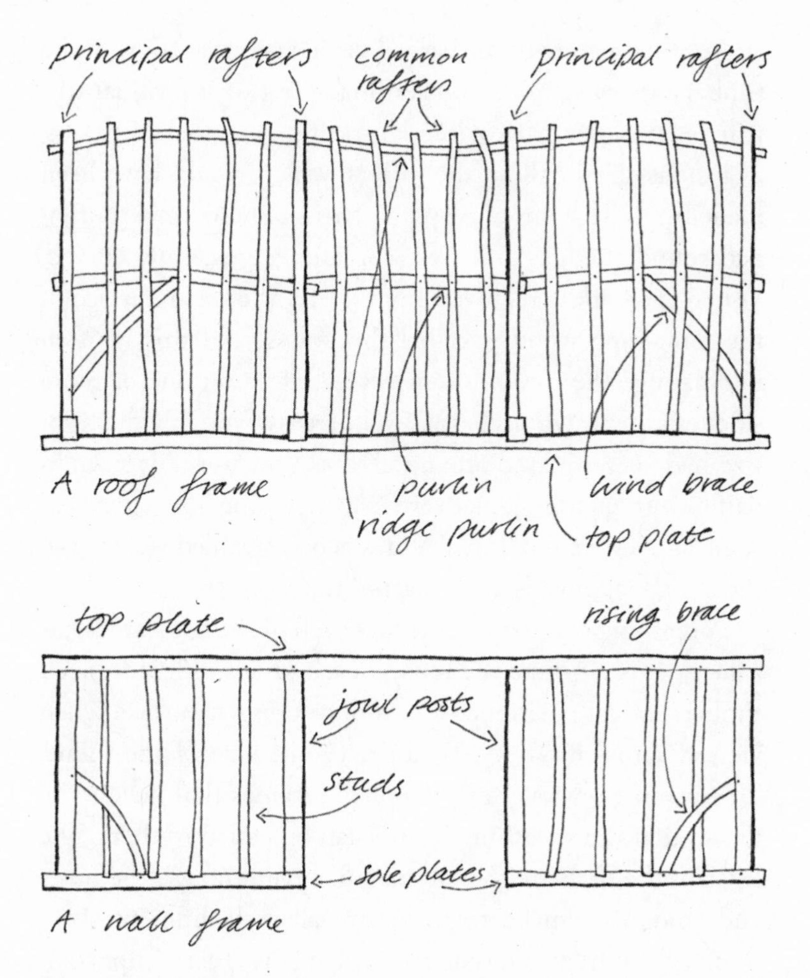


The components of the wall and roof frames.

strength was needed and where second-rate timber would work perfectly well. Significantly, they also shaped and cut their timber to follow the shape of the tree that it came from. As I will explain in the following chapters, there is a lot that can be done with a curving piece of wood that nowadays is set aside as forest waste.

Another feature of the barns I visited was that some of the timbers appeared to have been made out of individual tree trunks. This was true of the largest timbers, the tie beams, as well as the smallest, the rafters, which would have been made from the stems of very young trees. These timbers often had rounded edges and were sometimes gracefully curved. When the bark is removed from a tree, the outer surface of the remaining wood is called the 'wane', and this is often undulating. There were often waney edges on one or more sides of other sawn timbers as well. Trees of roughly the right size had been squared off and used as they were (or sawn in half or into quarters), inadvertently including any patches of wane as they went. Clearly, elm was being milled right up to the bark, maximising every part of the tree.

Something else to notice is that each of the flat frames has a 'face'. This is where the tops of each of the different-sized timbers are all positioned to produce one flat surface when viewed across it. They only align to one side of the frame. This is most obvious for the outside frames that have to be flat so that the sheathing boards can be nailed to them. The sides of the barn are therefore all flat even though the posts and studs that make them up are all very different sizes. Each of the frames has a face, and this is a very important reference when the carpentry begins. In my illustrations with the names of the timber parts, the faces are shown with a heavy line.

Having identified the face side of each frame, another feature becomes apparent. Numbers, known as carpenters' marks, are chiselled onto the face sides of many, if not all, of the main timbers. These numbers are commonly expressed in an approximate sort of Roman numeral system, and the regional variations are worth investigating as a form of vernacular numeracy. The carpenters' marks are labels. Each timber has a position in either a cross-frame or one of the bays between the cross-frames. All the timbers of the same cross-frame will have the same numbering, or at least some regular numbering system. This distinguishes which timbers go in which cross-frame. But it doesn't distinguish the front and back, for example, the jowl post on the front wall and that on the rear. For this there is a subtle difference in the markings. There is not a rigid universal system of what each numeral should look like or how front and back are differentiated. They may be made with a longer and shorter chisel, or one will be made with a curved gouge, or one numeral will have an extra diagonal arm, called a flag, added to one of the numerals. Some timbers may have more than one mark; for example, the external walls, to distinguish the bay number and the ordering of studs within it.

This variety in detail is part regional diversity, and is a kind of dialect within carpentry traditions. Keeping this diversity alive, along with any unique regional variation of framing design, is important. The carpenters' marks and the shapes of timbers in a frame are like the words in a language. They are functional and idiosyncratic at the same time. They give expression not just to the individual, but also to a process. They are part of what it means to be living in a particular place in a specific landscape.

From measuring the sizes of the timbers of local elm frames, a range of cross-sectional sizes became apparent. Obviously, the larger the barn, then the bigger the span or length of bay. Beyond that, there is also a small range in sizes, but for the barns in my region, a pattern emerged. There would be variation of an inch or so, but for a medium-sized wagon shed, which is what the Carley Barn approximates,

I could produce a list of the size of the timbers I needed. The tie beams were 9 inches deep by 8 inches wide, the top plates 6 inches deep by 7 inches wide, and so on. The smallest timbers were the studs in the tops of the gables at 2½ inches square. Everything else was somewhere in between.

When you stand in the middle of a timber barn, with great timbers high overhead, there is a thought that springs to mind. How did they get those heavy pieces of wood up there? Today, with cranes, it's not even a question that needs to be dwelt on. But before cranes, how was it done? Raising frames by hand is beyond our cultural memory in the UK, though there are a couple of classic American movies, *Seven Brides for Seven Brothers* and *Witness,* that have memorable barn-raising scenes. However, what is certain is that our forebears would have developed the detail of their frames so that raising them by hand would have required the least effort possible. More of this in later chapters. For the moment, whilst examining historic timber frames, it is worth noting that there is a primary frame of large timbers that is the essential structure. It holds everything upright and stops it wobbling. Then, secondary to this, are the studs, joists and rafters that the outer boarding, floorboards and roof covering are fixed to. By looking at the joints of the eighteenth-century barns that I inspected, I could see that these secondary timbers were sometimes added after the primary frame was raised. It may not be true for all frames or all local traditions, but I recognised it as a way of reducing the weight and number of timbers to be fitted together and simultaneously lifted during a frame raising.

With this appraisal of local barns complete, the framing drawings can be drawn. You don't need a computer for this,

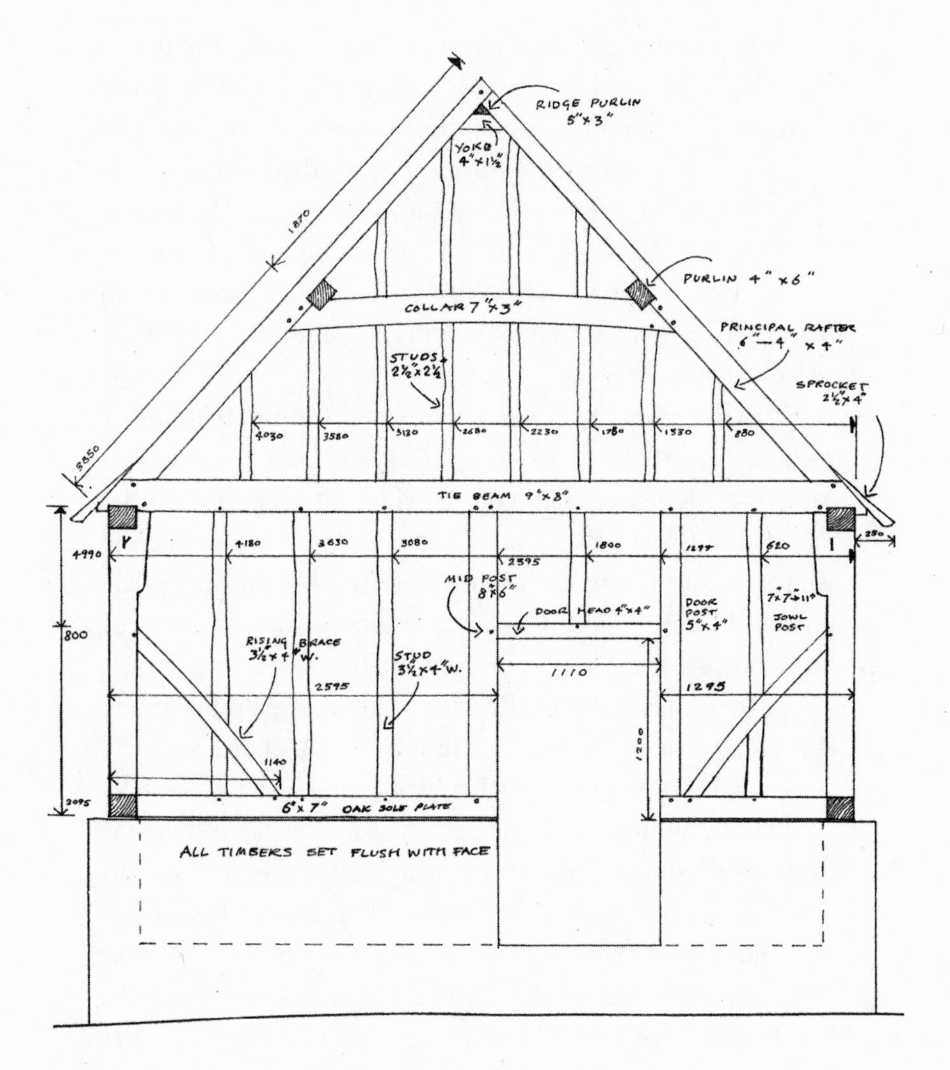


The framing drawing for cross-frame-I.

just a pencil, scale rule and paper, but there is drawing software for this if you feel you need it. Drawings of each of the frames should be made with the frame's face facing towards you and at a convenient scale, say 1:20. This gives enough detail to measure dimensions from the drawings. The drawings also have crucial measurements marked, such as those between the bay corner marks. So, the width of the barn, the length of each bay and the distance from the bottom of the sole plate to the top of the top plate are given. Along with the four outside walls, roof and cross-frames, there are the less obvious horizontal frames of the sole plates and the top plates. There may be horizontal floor frames as well. These should also be drawn.

With the framing drawings completed, a long list of timber components is drawn up: the cutting list. Each timber from each drawing is itemised with its width (as measured on the face of the frame), depth and length noted. To allow for the length of a tenon in each joint, an extra 4 inches (approximately 100mm) is added, or 8 inches (approximately 200mm) for timbers with joints at both ends. This gives a tight size. For peace of mind and to allow for a fault in the timber at one end, adding a generous further foot is not a bad idea.

At this point you may be wondering if this book is written in metric or imperial measurements. A curious feature of being a carpenter in England these days is that you end up thinking in both! For 50 years we have been decimalised in the UK. Yet, when it comes to handling dimensions of timber the approximate sizes of your body, hands and fingers, inches and feet are useful sizes and feel intuitive and tangible. An inch approximates to the width of a thumb. A foot to a foot. But this is obviously not true for everyone. My booted foot is reliably one foot long. My stride is one yard. An extended stride is 1 metre. The span of my hand, end of little finger to end of thumb is 10 inches. I use that one a lot. So, the point is more that you can use your own body as a reliable measure, once you have calibrated it and your whole system gets used to seeing the world that way.

Fractions again have a kind of bodily logic to them. The English version of the imperial scale on a ruler is divided into halves to give half inches, quarter inches and so on to eighths and sixteenths. Our eyes seem to be able to split these distances in half with relative ease, so it makes sense. Furthermore the use of dividers by carpenters, as the name suggests, allows division and multiplication of lengths through geometry rather than numeric mathematics. Your body, through your hands and the dividers, learns how to do the arithmetic, rather than your brain.

I imagine, as well, that it comes down to what you are used to. For me, dimensions are also easier to remember in feet and inches. Four foot six and a quarter is memorable amount, almost like a phrase. The ability to remember and transfer numbers walking from one piece of wood to another is limited after a long day's work. From experience, I would say that mistakes are more easily made when you think in metric! However, planning authorities in England require dimensions in metres. Timber merchants likewise supply timber lengths in metric. Plans and cutting lists are therefore usually in metres or millimetres. So, often you think about wood in imperial and write it down on drawings and lists in metric. I know, it's bonkers.

With all the framing drawings completed, the site itself can be marked out and a plinth wall built in advance of the frame. The laying out of the shape of the Carley Barn on the grass of the meadow by the large pond at Churchfield Farm was a big day in this story. It's a powerful moment to fix the position of a structure that will be in exactly that place for potentially hundreds of years. Esther, Bruce and I arranged to meet in their field to take the first step of making the plans become reality. Some disused electric fencing stakes were knocked into the ground to mark the four corners of the barn. We walked around the site to view the footprint from all angles. Attended by Clive, their beautiful lurcher dog, we stood back to try to imagine the barn completed in front of us. It is important to see how a building can most successfully respond to the site. We noticed the slope of the land, the relationship of our proposed positioning to the pond and to the fruit trees at the back. The corner stakes were moved slightly further from the pond and as close as we dared to the two giant lime trees that formed the backdrop to the site. One day the limes might blow over or drop a branch, so it wouldn't pay to be too close to them. The slope of the land also allowed the ground floor to drop down at the lower end and a two-storey bay with a storage loft to be included. With the stakes knocked firmly in place, the footprint was set. A shovel was produced and the first ceremonial turf was dug.

---

A timber frame needs to be kept off the ground and away from rising damp. This allows the sole plate to keep dry and free of fungal decay. So barns are placed on plinth walls. These can be stone or brick depending on local walling materials. In Hertfordshire there is very little natural stone suitable for this, but plenty of clay to make bricks from. So, traditionally, the plinths are made of brick. Ideally a plinth should be 18 inches tall to be out of the range of rain splashes. They can have a damp-proof course, though the old barns didn't have them. If the bricks are laid in lime mortar, moisture evaporates away and the sole plate stays dry.

Building the plinth was one of the first activities that brought the volunteers together. Many people hadn't met before. Some had more experience of practical work than others. Working as a team, standing and kneeling side by side, allowed people to get to know each other and chat as they worked. No one in the volunteer team was an experienced bricklayer, but that didn't matter. There was a range of capabilities, but we began to realise that the odd mistake was okay and everyone relaxed. The string lines stretched tightly between the corners gave enough of a guide to work to, and a bit of inconsistency in the finish gave a variety to complement the bricks themselves. No two bricks were exactly the same; they were originally handmade, and the firing left a range of colours, from red to a glowing orange.

Perfectly formed bricks laid in a monotone way would have been completely out of place in the barn. Just like the timbers that were once growing in a wood, the barn timbers will carry something of the individual natures of the trees, and this subtle variation will be complemented by the brickwork of the plinth. The same would eventually be true of the tiles on the roof. These were also reclaimed and handmade. Each had a slight variation in shape and colour that is difficult to imitate with machined tiles. Even the weatherboard cladding that was nailed onto the finished frame walls had a very slight wobble in its line. We used 7″ × ⅝″ oak, milled from a large oak tree being cleared from a garden nearby, and as the oak dried it produced a slightly uneven line to the edge of each board. In a building like a barn, that individual variation continues to be expressed, which is part of its beauty. It is the beauty of random imperfection that is always found in nature. Rather like the happy accidents of glaze patterns on Richard Batterham's pots, the random effects of nature, weather and innocent mistakes are reflected in the barns of countless village carpenters over the centuries. That barns retain the feeling of nature, of trees in the wood, is undoubtedly part of their enduring appeal.

FIVE

# A Walk in the Wood

The story of the carpentry of Carley Barn begins with a walk in the wood – Bushy Leys Spring wood, to be precise. This is the name of the small block of ancient woodland located about five miles from the smallholding where the new barn was to be built. The wood is not accessible to the public, so arrangements were made with the woodland manager and a date was set one Sunday in September for the Carley family and the Barn Club volunteers to see the trees that were to be felled.

A walk in a wood is a beautiful experience. To enter a wood really is to enter another world. So it was when we walked through Bushy Leys Spring wood, prospecting for trees. On that blissful late summer's afternoon, having walked across a meadow under clear blue skies, we entered the dark shadows of the wood. A wood has a big impact as soon as you are in it. Suddenly all the senses detect a whole new world. Compared to the visibility and openness of the fields and meadows outside, the enclosure concealed more than we could see. The canopy high above was contrasting dark and light, with leaves and branches speckled with bright patches of sky. There were the towering stems of trees and the smell of rich earth and undergrowth. The soil felt different under foot, so soft and yielding. Sounds also changed. There were close by sounds of our movements at the woodland edge, muffled by thickets of shrubs and the shrill alarm calls of birds resounding deeper and beyond where our eyes could see. It took a few minutes to adjust, to relax and take it all in.

One of the volunteers, Jo Sampson, whose interest in woodwork is matched by his interest in wildlife, explained how the sight of the elms took him back to his childhood. Approaching from the footpath, looking towards the far side of the wood, he was taken aback by the sight of the enormous and graceful trees:

> *On arrival, I couldn't believe my eyes: there before me, standing proud from the main canopy of the woodland, were the unmistakable shapes of a number of huge elm trees. The main trunks towered overhead. Seeing these lovely living trees, I felt quite emotional as they took me back to my childhood home, behind which were a number of very large elms like these.*

It was a poignant moment and one that he didn't think possible. These elms had been there for at least 120 years, yet Jo hadn't realised their existence in spite of living close by. He felt compelled to go and touch the trees and to simply spend the afternoon just being there. More people gathered under the elms as the afternoon progressed, appearing like explorers out of the undergrowth. I wanted everyone to have the opportunity to just spend time with the trees: observing them, being amongst them and gazing. It takes a while to slow down internally and to be happy just being still. Some, like Jo, just marvelled at the beauty of the trees, the wood and the wildlife we encountered. My wife, Lydia, and Bruce Carley examined the leaves and twigs, closely scrutinising the minute shapes and textures. Scarlett, our daughter, climbed a tree and stretched out along a branch, gazing up through the leaves. We absorbed the beauty of the place where the trees had grown, and we were all aware with a degree of trepidation that our role was to transfer that beauty and the presence of the trees to the building we were about to undertake. The sharing of the experience together was one of the moving aspects of the day and made it a lasting memory for everyone. It was where the journey of making the barn was to begin, and we felt we were at the start of an exciting adventure.

If ever there was an example of a whole being greater than the sum of its parts, it is a wood. Trees live well beyond our human lifespan and woodland, as an entity, even more so. A wood is a living window through which the past reaches out to us. In some rare places in the UK, and particularly in other countries, ancient woods are remnants of large primeval forests and a very tangible link to our planet in its natural state.

Something in your imagination engages immediately when you walk in woodland. As individuals I think we each make our own relationships with trees and woodlands, and forests as well, if we are lucky enough to spend time in them. Throughout our lives, from childhood to retirement, we experience trees. Most people can relate to having enjoyed the shade of a tree, lounging under the pattern of dappled leaves blowing in a breeze in the summer's heat, or marvelled at the iridescent leaf colour in autumn, seeing them almost glow from within. A track through a local wooded park or a path through an ancient wood, gives opportunity for bathing in fresh air and immersion in nature. Given the right conditions – maybe the overwhelming stillness, or maybe the sight of sunlight slanting through early morning mists – unique moments of connection can be felt that almost transcend into the sublime. When you sit in contemplation with your back against the trunk of a tree, you can only guess at what it must be like to be literally rooted to the spot for one hundred, two hundred or even a thousand years. These thoughts may be held merely as memories from childhood, but nevertheless, we all feel strongly protective when trees are threatened with felling and tropical forests are brutally destroyed. Wild forests as undisturbed natural habitats are rare on Earth, but they do still exist. Their beauty is breathtaking. For me, large forests are intimidating, dangerous and spiritual places all at the same time, and any visit to a true forest is a pilgrimage.

Bushy Leys Spring is a typical Hertfordshire wood and a mixture of different tree species: ash, hornbeam, oak, hazel and elm. I had noticed the fluffy tops and purple blossom of elms sticking out above the tall canopy during a previous spring, and I asked the woodland manager if I could take out any that had died or were beginning to turn. As a rule, I don't fell large healthy elm trees. I wait until they have died or are starting to die back, having just passed their prime.

There is an apparent contradiction between loving the standing trees in a living wood and wanting to fell them to use their timber. This tension should be held on to, kept alive and not suppressed. It's a really important feeling. It's one that has to be resolved in order to fell a tree, or indirectly cause the felling of trees by the purchase of timber products. It's important because human beings have clearly been responsible for the loss of millions of square miles of forests as populations grow. We have now reached the point where woods and forests need to be globally protected, or it is conceivable that, one by one, they will disappear. Our impacts on the Earth have grown to the current perilous situation such that we have to take responsibility for what we do, or what is done on our behalf. Through the transaction of buying a finished product that was made far away, we have very real impacts on wildlife and habitats, all of which we are unable to see.

In the past this wasn't so much the case. Our environmental impacts were tiny two thousand years ago, and nature absorbed our pollution and bounced back. But now we face mass species extinction and climate change at a global level because of human activity. Before the Middle Ages, our forebears may not have needed to be concerned about felling an individual tree in a large wood like the one that predated Bushy Leys Spring. There was no need. They could just concentrate on learning and celebrating their craft and making a living. But now I think we are each compelled to think more deeply about what we do and how we obtain material resources. This is a practical necessity, but also an opportunity to reach deeper and to redefine our relationship with materials and the natural world. In this respect, using local resources is good, because you can immediately see whether your actions are unsustainably destructive.

When I fell a tree, I spend a little time in advance looking at the tree, appreciating its beauty in many ways. I try and really 'see' the tree. To pause in this way is a kind of playing out of a negotiation whereby I hold in mind what I am trying to make and the fitness of the tree in front of me for that purpose. It is intense, and I often hold the image of the tree in my mind's eye for many years after felling. Sometimes, very rarely, I will leave the tree un-felled and walk away. The moment before making the first cut with the chainsaw is also very poignant. As you lean on the tree to decide the line and level of the wedge of timber to be removed, there is a moment of power over the fate of that giant being. It is a dangerous moment for the forester, if it falls the wrong way. And it is a point at which the tree is about to alter state. The chainsaw bar sinks into the wood and you are committed. There is no going back. And as the tree is felled and crashes to the ground, all the potential of that tree's wood, its timber, is suddenly available to create something equally beautiful… or not.

The truth that we humans can be both destroyers and creators is self-evident. I think immersion in nature and acceptance of the cycle of life and death is part of how to balance these conflicting truths. Clearly, trees, along with other forms of life, regenerate. In nature, death is not final but part of a cycle. In a healthy wood young trees spring up where a clearing has been formed. As human individuals we tend to see death in highly personal terms, if not something completely final. But the more time I spend working outside, sometimes feeling strongly connected to the natural world around me, the more I feel part of my surroundings. After spending a length of time in natural environments, a strange sensation arises when you don't really distinguish where the 'you' of you ends and where your surroundings start. You feel less of an individual. I feel in those moments of connection as part of a more universal state of being alive, surrounded by a far larger flux of life, in all its other forms. Being part of nature in this way is a fleeting moment of deep knowledge, but the memory holds on to it. Life, and the experience of being immersed in living systems, feels like being surrounded by fully integrated beauty. Momentarily catching a glimpse of this web of life is what I think of as a transcendental experience, and it takes your breath away. Nature gives and regenerates relentlessly. An individual might die, but life lives on.

From the standpoint of these thoughts, it is clear that in natural cycles there is a harvest that we can take that does not decimate the resource or negate harvests in the future. Of course, given modern technology, it is easily possible to fell trees and spray chemicals in such a way as to destroy a wood completely. However, it is just as possible to fell trees in such a way that the wood will not be destroyed. Nature will respond with new, vigorous life. In this way we can take what we need but hold back from taking everything. It's less materially efficient to leave some trees un-felled, but they will reseed the spaces that are made. More than that, I believe we should also leave some of the biggest and most beautiful trees, not just because they provide a rich habitat, but simply because they exist.

Fully mature trees are a glory of nature, and the human experience of the world is diminished when they disappear. Foresters who felled elms after the outbreak of Dutch elm disease in the 1970s experienced the scale of these trees. One enormous elm proved to be hollow when felled and so large that a Mini Cooper car could have driven inside. Like other elms, the tree had developed a rotten core at some point in its life, but had sealed the cavity and continued to grow until Dutch elm disease had killed it. These were truly awesome trees.

In the UK we could collectively decide to fell fewer trees than we have in the last century, to resist felling when they reach seventy years old and let the best grow on to become giants. To balance that shortfall, we would need to make better use of those trees that were felled. At the moment we are doing the opposite. In the region where I live the vast majority of trees, including large ones, are being felled in the local woods and chipped by giant machines to be dried in heated sheds and burnt as biomass to generate green electricity. Fewer and fewer local trees end up at a sawmill and the timber is not being used at all. In this respect we are sending thousands of tons of perfectly good timber up in smoke. In the past we would have used these trees to make useful things. This is a challenge that faces our era and we will need ingenuity to reverse the trend. If anything, this calls for more depth of public understanding of trees, of wood, and a deepening of our crafts and wood culture.

The forester that was to lead the felling of the elm trees for the Carley Barn was Andy Mason. He has a love for wood and owns a small timber yard and log business near Cambridge. He also runs a sawmill there to plank the best trees that come his way. Many of his trees would otherwise be sold as firewood, so he is bucking the trend. He deliberately encourages his customers to come for just the odd board or two, for a DIY alcove shelf or something made from local timber. He wants to share his love for timber, and people can see the logs, the mill and the drying stacks for themselves.

Bushy Leys Spring is a working wood that has produced timber sustainably for generations. The 'spring' part of Bushy Leys Spring seems to suggest a source of water, though the word here has an ancient use and is a reference to the wood being a coppice wood. When deciduous trees are felled close to the ground, the tree base, or 'stool', regrows with fresh sprouts. These in turn grow into numerous stems, or, if thinned out, will produce single trees. The stools are closely grouped to encourage tall, straight regrowth. These new stems will have the full power of the roots of the stool to push them up fast and strong, much more quickly than a fresh tree seedling. Although true for elm coppice stools, it is also the case that elm will spread via new root shoot suckers, powered upwards by the nutrients from the parent tree, just waiting for the signal of fresh clear sky above to begin to spring upwards.

Coppicing is an ancient system of managing woodland, and it has been used in many woods in England during the past thousand years, though the majority of coppices are now overgrown and gangly, becoming 'overstood'. This neglect causes problems of over-competition between potentially large trees all of a similar age and none with a large enough crown that would normally support a tall tree and provide it with the energy to keep growing healthily. Any trees that become stressed are naturally more susceptible to fungal damage and being blown down. I have noticed that overstood coppice trees like this often develop heart rot – a fungal disease – at their bases and are eventually thrown down by winds.

By counting the annual growth rings of the felled elms at Bushy Leys Spring, I calculated that the last time that coppicing had occurred was around the year 1900. It is possible that some of these elms were beginning to be affected by Dutch elm disease, though the number of trees with heart rot at their bases may indicate they died of other fungal infections. Some treetops had been snapped off in strong winds, with lots of new twigs growing sideways in all directions from the stunted stems. Normally, a commercial sawmill would not take trees in this sort of condition, or anything like it. These stems would be firewood at best. Rotten, or 'doated', wood is not structurally sound and would be no use to me, either. Doated elm has a light cream, slightly bland colour, often with thin white lines and dots running across the grain. This degree of decay seems to form when the tree has been dead for a while, or when it has been left lying on the ground. Up until the early twentieth century, unsound timber that was beginning to rot was traded as 'wrack'. Conversely, I have also found dead standing elm trees that have simply dried out to become hard, seasoned wood rather than wrack, so it is not always easy to tell how rotten and unusable a dead tree trunk will be. A timber buyers' guide from 1910 states that timber that is milled from dead wood, though not necessarily unsound, was termed 'torrak'. Therefore, most of what came out of Bushy Leys Spring was torrak. It is interesting that as recently as one hundred years ago millers would consider buying and using dead trees and even had a word for them.

---

As you spend time in woodland you get to notice more and more the different smells and sounds of the different types of trees. Near where I live there is a large wood with very different sections or compartments. The Corsican pines that you first encounter when you enter the wood have a strong fresh pine scent, whatever the time of year. When the wind blows strongly, the roar is constant, a rasping unsettling sound like sucking in breath between your teeth. Hornbeam, by contrast, creates a deeper note, a fluctuating and rushing sound, having leaves rather than pine needles. Of course, winter winds in deciduous woods, when the leaves have gone, sound different again. The woodland floor under sweet chestnut coppice smells the most distinctively rich and earthy, whereas elm woods hardly smell at all. Unlike other trees, the leaves of elm rot away very quickly.

There are different sorts of ground flora in elm woods as well. The elm wood at Boxworth in Cambridgeshire has a complete carpet of cow parsley beneath the trees, and little else except ivy. In the spring there is a heady, acrid scent. A wood cannot be instantly assessed, and the matrix of plants, trees, soil, birds and animals has to be understood over time. Some creatures, like us, are just woodland visitors. One of Bruce Carley's favourite memories of the wood was seeing an impressively large fallow deer stag slowly trotting away across the adjacent field of winter barley, moving with complete confidence and regal composure, his crown of antlers held high.

We each personally perceive and engage, to a greater or lesser extent, with the woodland where we live. As someone who uses wood for a living, I suppose I am more dependent than most. Historically, sawmill owners would have held a deep understanding of their local woodland and what could be harvested sustainably. Every part of each wood would have been known. Some parts of the woodland would produce sweet chestnut good for cleaving into fencing, some not so good due to spiral grain. Some oak good for joinery, some not so good due to heart shake. Earlier in my career I could draw on that knowledge when buying timber. Sadly, this is no longer the case.

Last year I visited my local sawmill at Ayot Green. It's the local mill for my village, and there has been a sawmill there for donkey's years. I needed some oak and western red cedar. So I did what people have done for generations and approached the local supplier. The mill is medium-sized, always busy and has a great pile of tree trunks stacked in the yard. Finding the enquiries office, I asked the manager if they had timber milled from Hertfordshire trees. The answer was no. I asked if they had timber from trees grown in the UK. Again, no. So where do the trees he mills come from? He shrugged his shoulders and pointed to the computers in the office behind him. 'It's a commodity,' he said. 'I just get the best deal. I don't need to know where it is from.' Needless to say, I left empty-handed.

As recently as ten years ago, it was possible to buy locally grown timber, sawn in the local mill. So, something fundamental has changed in as little as one generation. We have transitioned from the local mills knowing where to source, for example, beam oak from the surrounding landscape, to a system that cannot necessarily identify country of origin.

Arguably, timber all looks the same once it's sawn up. Anonymity is inevitable. Whereas, knowing where a particular tree grew and following its journey to become a beam or part of a new wooden floor, is a story that belongs to the piece of wood. These stories are part of the culture of connection to nature, and that connection can be heartfelt. In spite of globalisation, there is a growing demand for a kind of product with a strong sense of provenance and the interest of Barn Club volunteers to visit the woodland that produced the timber is part of this search for authenticity.

Some of the elms in our wood were enormous and they had a different story to the rest. The one with the largest diameter was still healthy. It was older than the rest but not as tall and was completely invisible from the road, just 200 yards away. It had great sideways spreading limbs and a hollow halfway up with hornets flying in and out. Its deeply fissured bark and gnarled trunk, 4 feet in diameter I would guess, supported a majestic spread of branches. This indicates it was already a developed tree when the coppice was last cut, probably kept as a shade tree to protect the delicate early regrowth from the coppice stools. Its wide spread was in stark contrast to the very narrow trees with straight stems and few lower branches that were the coppice trees, competing with each other as they grew towards light and air. Some of these coppiced trees were the tallest ones I had noticed from the road. When a tree gets to 90 feet tall or more and has a clear stem for the first 50 feet, you really feel that you are in the company of a giant. You feel compelled to walk right up to it and gaze upwards. They have such a strong presence, and you can only feel but a dot in comparison. I think this is an important stage of the process of making a timber structure; to fully appreciate the timber as individual trees existing at their source: the wood.

---

As a craftsman today, connection with the source of my materials is a way out of the predicament of the anonymity of timber and complex modern supply chains. Awareness of the impacts, immediate and long-term, of harvesting trees for a particular project is a responsibility that I have to bear in mind. For example, I feel confident that coppicing a hazel stool to harvest some runner bean poles from the wood down the lane is okay. The pole isn't durable, but will last a few seasons in the kitchen garden. Stout poles of more durable bamboo are available in garden centres. Where are they from? Not from England, for sure. Probably the Far East, where it might be native, or perhaps elsewhere in a new plantation. Long-distance transport entails carbon dioxide emissions that we can no longer withstand if we are to reduce carbon dioxide emission. Was the bamboo plantation created by replacing naturally mixed biodiverse forest with a monoculture? Was bamboo even native to the country? It is impossible to say. So, I just use hazel rods instead. For any person buying a product, long supply chains are inherently more complicated to assess for sustainability. Knowing what degree of moderation was applied to prevent over-extraction in a far-distant forest is crucial. Restraint is needed.

Eighteenth-century Germany had an expression for this. In response to an impending timber shortage, it was decided to survey the German forests and develop a system of forest management. The word and the concept that a forest surveyor, Hans Carl von Carlowitz, invented in 1713 was *Nachhaltigkeit.* This was the first use of the term 'sustainability', but it more literally translates as 'holding-back-ness'. Clearly, if all the good trees were felled at once, and there were no more endless wild forest to extract from, there would be trouble ahead. He also coined the term *Holtzsparkünste,* the 'art of conserving timber' through maximising the use of each tree felled, which is the other side of the coin. We could learn from this one as well. These two principles, restraint and thrift, still hold good today and should guide us in the management of woodland and the use of wood.

Whatever timber we take, therefore, has to be used with consideration of its inherent qualities, and with a view to honour its potential. That is why just felling trees to be burnt for biomass is instinctively wrong when so much else could be done with the wood. When I initially looked at the dead and dying elms with the woodland manager at Bushy Leys Spring, their fate was to become biomass or be left to fall and rot. I explained that I would mill them and make the components for a timber frame that could last for centuries. He repeatedly stated, 'This is what we should be doing, this is what we should be doing,' and he was right. In reality, beautiful English landscapes hold an abundance of materials we can use. Barn Club is a way of immersing ourselves in nature and engaging fruitfully with the landscape.

Having meandered our way through the outskirts of oak and hornbeam at Bushy Leys Spring, we gazed at the interlocking shapes in the canopy overhead. Without warning, a hare started up and was off and away before we could react. I'd always thought of hares as animals of the open fields, but it is surprising how often you see them in woodland. After a while exploring, we gathered in the area where the elms grew. There was a mixture of healthy, dying and dead trees. I recognised the type of elm. They had small, pointed smooth leaves. The twigs were thin and had an open structure. They already had the tight buds of the flowers that would burst out in the late winter. The mature branches swept down and the bark was rough, furrowed, and formed lattice patterns. This tree is known as East Anglian elm. This type of elm grows to a good age and size, and can be found across Hertfordshire and surrounding counties.

I talked with the Carleys about which trees would be the right ones to fell. We looked at the shapes of the tree trunks and branches to imagine which part of the eventual timber frame those shapes could become. Slow curves at the bases of wide trees indicated arch braces, straight lengths between branches looked hopeful as top plates and principal rafters. The flared bases of the larger trees looked ideal for milling the jowl posts, which are the main supporting posts of the barn and carry the most complex parts of the timber frame. Smaller lengths and smaller trees could be studs, rafters and braces. There wasn't enough elm for the entire frame, but my gut feeling was that the primary timbers were all there, inside the trees standing in front of us. Apart from gauging how each tree could be used, we also had to consider how the trees would be felled without getting snagged in each other, and the route to the edge of the wood if the logs were not to

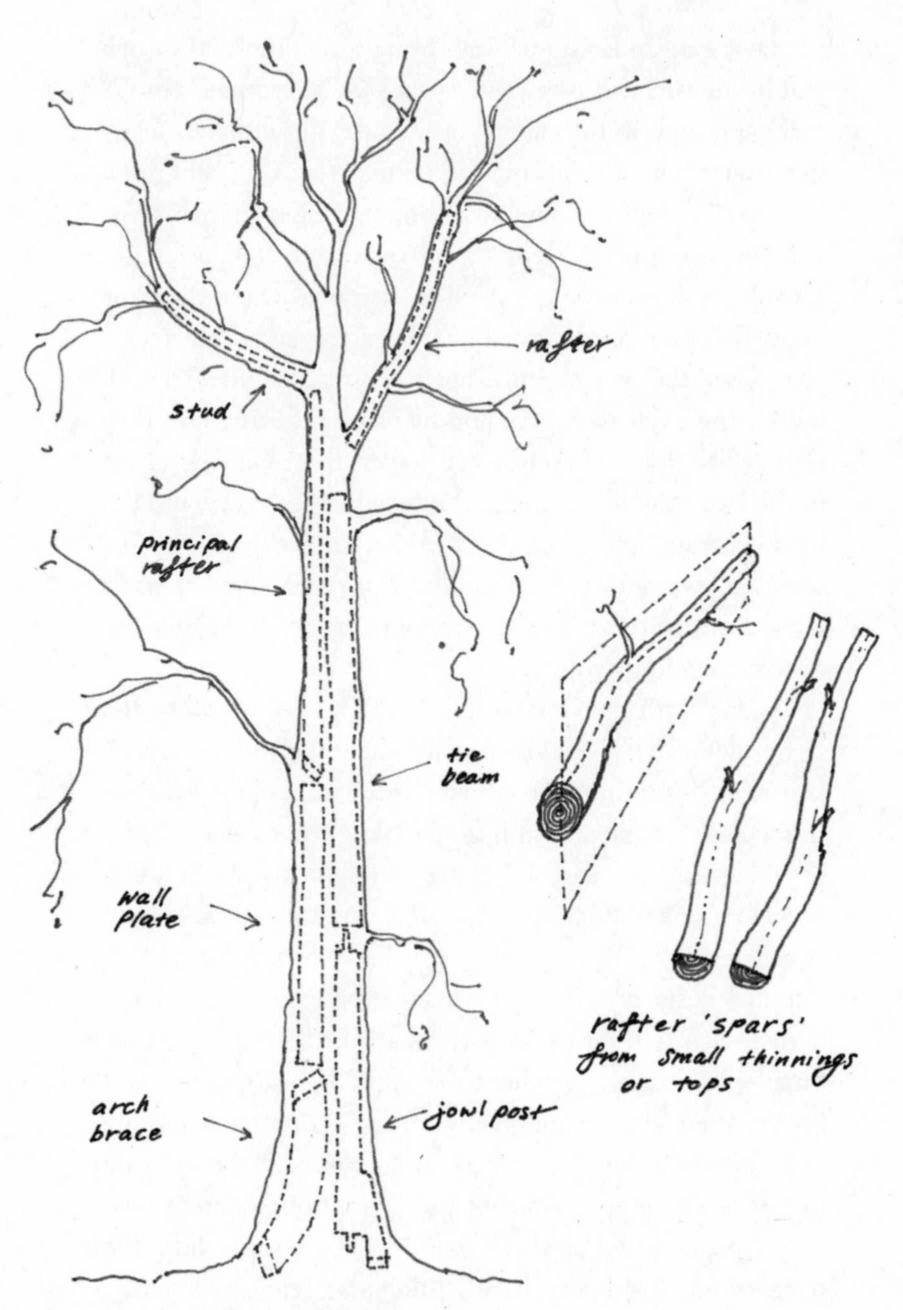


Seeing the timber shapes within a tree.

create any damage in their extraction. Eighteen blue crosses were painted on eighteen tree trunks ready for the forester to view and decide his plan of action.

As we pondered the timber that these trees would yield, a buzzard flew in and sat on the topmost branches of the tallest tree. From far below we posed no threat, but possibly sparked curiosity in the bird. We all stopped to look up at it. I love this part of a project. To find the timber you will need by going direct to the trees is a rare thing in today's world. For most timber framers it is a matter of emailing a cutting list to a timber merchant. But to go and search for the right trees in the midst of an ancient wood, with history in all its shadows and its branches teeming with life, is a process that feels incredibly rewarding. I am always keen to share this part of the work with clients, and the Carleys and their daughters as well as the Barn Club members were all moved by the significance of this part of the story of the barn.

I set a date with Andy Mason and his team to come and fell the trees. There was just a brief window of time to fell, bring them forward to the edge of the wood and deliver them to the site. The hunting season had started and the wood was part of the local game shoot. Elm was traditionally felled in November. I imagine this is because the tree is then most dormant and the sap content of the wood is lowest. Sap provides the food source for woodworm and fungus, so timber that is summer-felled is more prone to infestation. Elm is milled right to the bark, so sapwood will be included in the timber frame. To minimise the loss to fungus and woodworm, the timing of felling is obviously important. Added to this, in Hertfordshire, the autumns are dry until about November. After then the rains arrive, the temperatures cool and the clay soils become soft. The land becomes less passable to heavy vehicles without making deep ruts, whereas we wanted to minimise any impact on the ground and woodland understory. The last week in October was set for the felling.

Extracting the timber using horses would have been a great experience, but too slow for our timeframe. Instead, I decided to use a modern set of big machinery and do the job quickly. Andy Mason provided the heavy equipment and two tree surgeons, the Raymond-Tarplee brothers Rowan and Shannon, to work the chainsaws. A tractor with a long rope would pull the falling stems in the direction we wanted to fell them. This helped minimise damage to the remaining trees. A powerful swing-shovel excavator would lift the elm stems to a meadow at the edge of the wood, and its wide tracks would hardly leave a trace on the woodland floor.

In the past, trees were hauled out of the wood by dragging, known as 'tushing', or by using a timber arch to raise the log off the ground and hold it suspended between a pair of large wooden wheels. The trees would be cut with the leading stem, the spire, in place and then transported in the longest lengths possible. Horses were the source of power for this. Instead of an excavator or powered timber grab to load a trailer in a clear lift, the logs were rolled onto the wagon sideways, called 'parbuckling'. This is done by bringing the wagon parallel with the log and passing chains over the wagon, under the stem at each end, round and back over the top, and back across the wagon. By pulling the chains the log rolls up ramps onto the wagon. Mechanically, it is a very efficient way to lift long, heavy logs and keep their full lengths. I have used this technique to move a large log and it is surprisingly easy to do. After parbuckling, the trees themselves became the chassis of the wagon, supported at each end by a pair of wheels. It seems extraordinary to us now, but it would have been normal in the past for much longer trees to be transported than today. The distance to the carpenters' shop would also have been much less than the distances that timber trucks cover, and carpenters would mill the timber to the lengths they needed in their own yards.

After felling and 'forwarding' to the edge of the wood, each full-length elm tree was laid out on the meadow grass. When it came to transporting the trees to Churchfield Farm, the site of the Carley Barn, I was not to have the luxury of having each tree trunk kept intact. They had to be cut to the shortest length I could get away with to fit on the transport trailer. Each long stem was assessed to confirm the size, shape and length of timber that could be milled. If all goes well, every timber from the cutting list is accounted for. By having calculated every single timber I needed for the timber frame in advance, I was able to measure tight sizes and know the minimum I could afford to cut to.

To use the most of each stem in its straight trunk before its first branch, there is a bit of juggling around of potential lengths. Once into the crown, the stems become more winding and the side branches leave large knots when milled, which will weaken the structure of the frame. Sometimes it is possible to mark a length above and below where a large limb appears. This is stressful work because once a stem is sawn there is no going back. I had a cutting list on a clipboard, but with over three hundred different pieces of wood on the list, there is not a lot of time to assess each log for size, curvature, quality and combinations of various lengths available. Added to this, with elm you sometimes find complete birds' nests hidden and sealed within a tree. I had read about this, but had not experienced it until sawing open these elms. We did indeed find a bird's nest within a hollow void inside one of the stems. It was lined with white horse hair and had two small blue eggs, and was completely encased within two inches of solid wood all round. A branch must have been damaged, fallen off and rot established and locally hollowed the tree. Then the living tissue of the tree, its cambium, grew back over the years and sealed the hollow, nest and all. I did not know it at the time, but there was another bird's nest in another hollow in one of the other logs. The log looked perfectly intact from the outside and was assigned to become a tie beam – one of the primary timbers of the frame. This nest will appear later on in the story.

After an intense half hour of assessment, the elm trees were marked and chainsawed to the chosen lengths, with a foot or so extra where possible. In cutting to length it would have been wonderful if all the trees gave long lengths of regular prime wood, but as a small clearance felling of trees from neglected woodland, that was never going to be the case. It was a typical mixed bag of shapes, lengths and sizes. Over the years I have learnt to prioritise the tie beams, top plates, principal rafters and jowl posts. It seems that then the rest can be sawn without generating many surplus lengths or sizes. It always surprises me how efficient this approach is. By working to a cutting list of shapes to produce a traditional elm barn frame with its capacity for inclusion of curving timbers, most of each tree can be used with nothing much surplus or wasted. In all, our forestry operation worked very well and we managed to fell, extract and transport the trees in two days. We also ended up with two trees that were wide enough to provide sawlogs for milling into floorboards. By calculation and with a measure of gut feeling, our 18 trees, ranging from 10-inch to 30-inch diameter, gave us 70 logs, weighing about 25 tons in total.

These were loaded on a timber trailer and took the short journey to Churchfield Farm to be stacked in the yard. The heavy elm logs were lifted off with ease by the hydraulic timber grab and placed with delicate accuracy on bearers to rest in piles of similar length.

To fell and move these 120-year-old trees so efficiently felt both empowering and a bit frightening. It was easy to see how an ancient wood could be destroyed if we had taken things a little further and felled a few more trees. As it was, the impact on the woodland was minimal. We just felled the trees that we needed. A glade had been formed, something that a strong winter storm could have achieved, or the death and eventual collapse of the aging trees. Light would now penetrate to the woodland floor and dormant species of flowers such as foxgloves would appear from seeds kept safe in the leaf litter since the last time. Young light-suppressed saplings would also surge upwards, along with the regrowth from the stools. But we all felt in our hearts that feeling of shock of the sudden felling of the tall trees long after the sound of the crash as they hit the ground.

Esther Carley suggested we return a few days later to spend time in the wood, to tidy up if we needed to and to stack some of the twiggy brush around the coppice stools to protect the new shoots from grazing deer and rabbits. It was a good excuse to spend more time in the wood, with the promise of a cup of tea brewed on a Kelly Kettle using dry twigs from the ground. With Chris Whitehouse from Barn Club, we returned a week later and spent an afternoon setting the understory to rights in the dappled sunshine of the wood, enjoying the new glade for what it now was: a part of the wood ready to re-establish and become a new chapter in the life of Bushy Leys Spring wood.

I returned to the wood exactly two years later. There had been an explosion of growth in the meantime. At this point in the wood's cycle, brambles were dominating and making passage by foot impossible. The glade was thick with their barbed stems. Deer and rabbits would equally struggle to walk through the tangle to nibble at any saplings. This is the messy stage in the coppice wood cycle, though when the tree canopy grows back across, the brambles will be starved of light and die back. The purpose of my visit was to see how well the elm coppice stools were doing. They were thriving. Elm shoots were poking up through the brambles. The largest coppice stool, which had supported the 90-foot-tall elm, had already grown to about 15 feet tall. It was fantastic to see the regeneration of the elms so successfully underway. The woodland was replenishing all by itself.

SIX

# The Red Elm, the White and the Sand

The moment the blade of the sawmill makes the first pass through the log is a moment of truth. All the volunteers gather round to look as the first slab is lifted off. It's a captivating moment. There is always an air of anticipation as the mill's blade lifts clear and the top slab is turned over. After all the effort of finding the trees, getting them felled and extracted from the wood and transported to the site, the moment arrives to cut the first slice and all eyes are looking to see what the grain is like. What about colour? Are there any knots or hidden defects? These questions are answered in seconds as the sawdust is swept away with the glove of a hand. This chapter examines the wood itself, how it grows and the particular physical qualities that make it so beautiful and that have made it such a useful timber over the centuries.

Elm is a tree of secrets, so it is no surprise that the wood is the same. Usually, freshly cut elm is a deep orange colour, but sometimes there are flares of purple and pink or lines of green and dark brown. There are swathes of lighter and darker tones and usually a ring of paler sapwood immediately under the bark. All these tints are the colours of freshly sawn wood, looking as bright as a freshly cut watermelon. As the timber reacts with oxygen in the air and begins to dry, that bright, watery vigour of the immediate colour begins to fade within minutes or to disappear completely. The orange turns to brown and the pale sapwood eventually darkens to match it. The surface evens out to more of a honey-brown, though some streaks of colour remain, including an occasional distinct green line. But, as it dries, the pattern, or the 'figure', of the grain also becomes much more visible.

The figuring on some pieces of elm is phenomenal. It seems almost tropical in richness and unlike any other of our native trees. Describing the grain as flowing water is particularly apt. The lines run together tightly at one place, then pull further apart, only to draw together a few inches further along. It is as if the grain is rolling, churning over itself. It eddies and swirls this way and that, as if directed by unseen forces beneath the surface, like water over boulders in a river. When the timber is rolled over, there may be thousands of tiny small flecks of darker brown across one face, not unlike a piece of beech wood. On the other hand, sometimes elm will be milled to reveal a steady straight grain with no particular movement. It depends on the variety of elm and its growing conditions.

The range of elm types in the East of England is reflected in the diverse appearance and quality of the timber. Even elm types that look similar can produce timber with very different characteristics. The most expressive and swirling elm grain and by far the toughest elm that I have found from any source comes from some large elms in the hedges of just two fields of a farm near Royston, Hertfordshire. According to the farmer, the timber from the elms growing in adjacent hedges, although looking similar as living trees, was completely different and straight-grained by comparison. I have read that village carpenters knew exactly where to go for elm wood with particular toughness, and from my experience with the swirly-grained elm, I can understand why.

However, within all the variety of appearance of the sawn wood, there is always a big clue that lets you know you are looking at elm wood. Between the curves of the rings of yearly growth is a particular pattern. A myriad of minute arcs zigzag between the rings, with perhaps as many as fifteen layers of this pattern between each clearly defined annual ring. This figuring is seemingly random, but is consistent in every piece of elm wood and adds enormously to its beauty. When the wood is seasoned and sanded, the effect is more pronounced and spectacular. Many old chair seats were made of elm, identifiable through this partridge-breast figure, and its beauty only improves with polish and use.

Elm is the only native British tree that does this, so why is it different? By looking at the biological structure of a broadleaved tree, we can understand what is going on at a cellular level to produce the physical qualities of elm wood. And, importantly, its suitability for making a barn. Elm is a deciduous, broadleaved hardwood tree, as opposed to an evergreen softwood conifer like Scots pine. Hardwood trees have formed the majority of our native woods in England since human beings first arrived here and started managing woodland and using timber. Since then our landscape has been dominated by oak, elm and ash. Conifers, on the other hand, have become the staple diet of the modern timber industry, and these were only extensively planted in the last two centuries.

The structure of a deciduous tree is sophisticated and beautiful. Of course, there is the famous photosynthesis in the leaves that produces glucose and oxygen from carbon dioxide and water. That would be impressive enough, but there is much more going on out of sight behind the bark. There is a very thin and extraordinary sleeve of living, growing cells just below the surface. This is the cambium. It covers every branch, twig, root and stem in a continuous layer, like a pair of tights. It is the part of the tree that grows. Contrary to common sense, rather than outwards, the cambium grows inwards, pushing itself away and forcing the sleeve to stretch, growing bigger and outwards as it does. In this way it always stays on the outside layer of wood. This cambium layer is completely at rest in the winter and wakes up in the spring to start growing new cells. The tradition of wassailing in the orchards in January to 'wake up the trees' is more literal than you might imagine: depending on the species of tree, it is some time in the winter months that the sap begins to rise to feed the cambium. Sap is the blood system of the tree, and although consisting of water, within it are dissolved sugars, minerals, oxygen and everything that keeps the tree alive.

On the inside surface of the cambium sleeve, three types of cells are produced as the tree awakens. The first type form vessels made of lignin, called 'xylem', through which sap from the roots rises to reach the leaves as they appear above.

In the spring, a new ring of tightly packed vessels are therefore formed just inside the cambium, termed 'early wood'. The vessels are large in the case of elm, so large that they can be seen as dots with the naked eye in the cross section of a log. As the spring progresses, a second type of cell forms fibres made of cellulose produced by the cambium, mixed with new smaller vessels overlaying the early wood. During this phase the strong and flexible fibres predominate. These hold the tree up, whilst their elasticity allows them to sway in strong winds. This is termed 'late wood'. They also run longitudinally in the direction of the stem and branches.

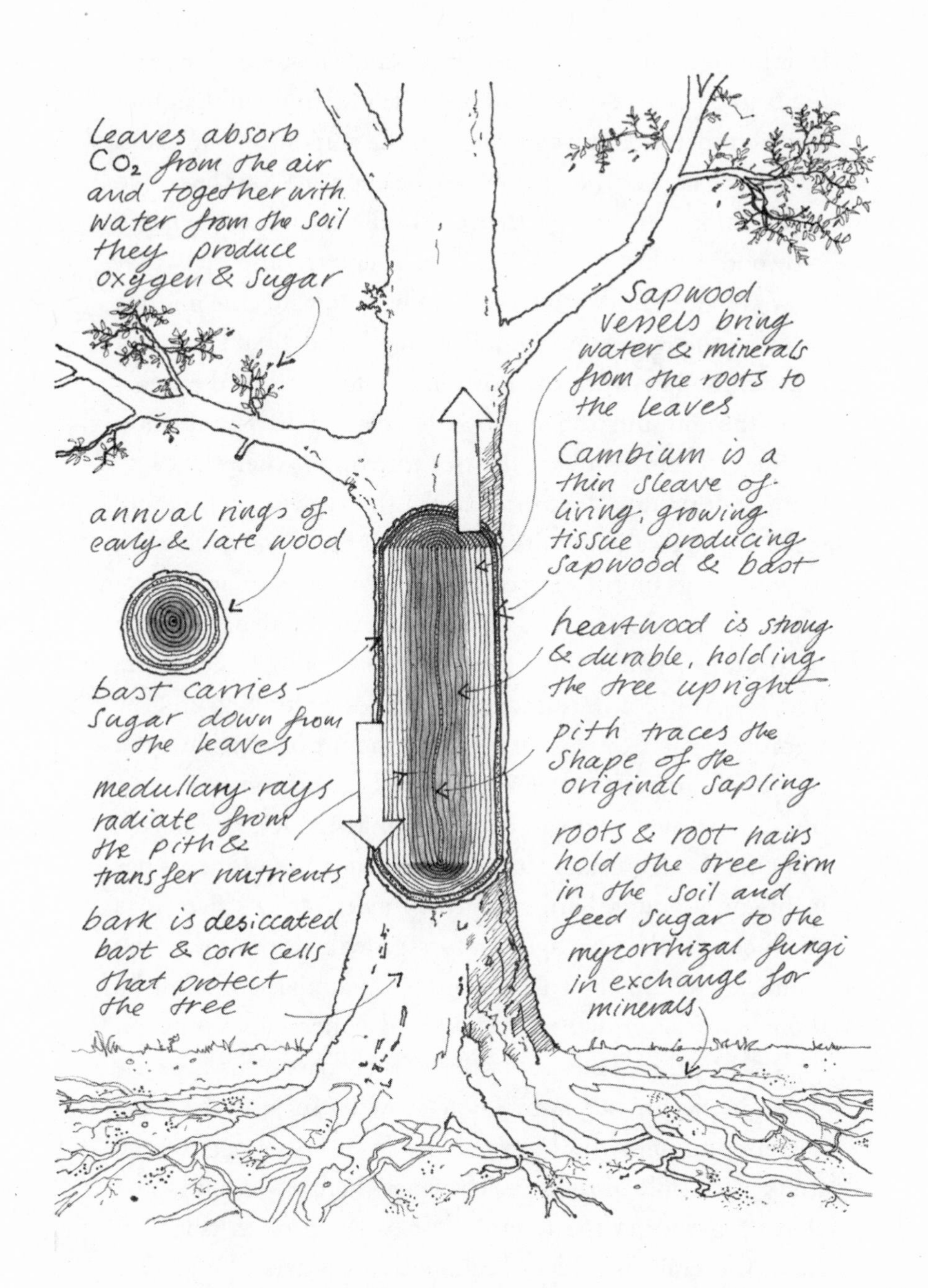


Component parts of a broadleaved tree.

Interestingly, most fibres also grow in a slight spiral around the trunk. An area of root on one side of a tree can therefore spread minerals and water through the spiralling to all sides of the tree above. The sap always rises upwards on the inside of the cambium through the vessel cells and gives the name 'sapwood' to this part of the tree. There are also cells that grow radially, or sideways, reaching back towards the middle, or pith. These are the 'medullary rays', and they store and transfer food between the inner and outer parts of the tree.

In the autumn the process of creating more sapwood declines to zero, readying the tree for winter when the cambium is dormant. The difference in the types of cells being made as the tree grows fits exactly with the cycle of the year, so these rings of differing cell types create successive annual rings. These rings are visible when a tree is felled. As every child knows, count the rings and you have the age of the tree. The width of the annual rings varies depending on how vigorously the tree is growing, the amount of sunshine, the conditions of soil and climate, as well as on the species of tree. The annual ring can be as much as ½ inch in oak, ash and elm, though usually it is less. It is measured in years per inch, and a young hardwood tree might achieve five rings per inch. As the tree ages and its growth slows down, the number goes up to ten or more. The annual rings can be thought of as growing in elongated cone shapes, one on top of the other. Trees are wider at their base. This is partly through the greater number of growth rings at the base, but also through the buttressing of the lower part of the trunk – the bole of the tree. The rings themselves tend to be closer together at the treetop. When the growth rings are close, the grain is said to be finer and easier, or 'milder' to work with.

As the cambium expands further and further outwards, the new growth overlays the rings made in previous years. After several years the older rings become redundant for sap transportation and die. The number of years this takes to happen varies amongst species. For sweet chestnut, this may be just three years. For others, like oak, it is around fifteen years. Thickness of sapwood varies between species and growing conditions of the tree. As they die, the sapwood cells convert to heartwood. A chemical change occurs in these inactive sapwood cells that gives them another role in the structure of the tree. The once porous vessels in some species like oak become blocked with minute knobs called 'tyloses'. These heartwood cells are now waterproof, durable and strong. Oak and sweet chestnut trees also generate natural preservatives called tannins that flood the heartwood and make these timbers particularly rot- and insect-resistant.

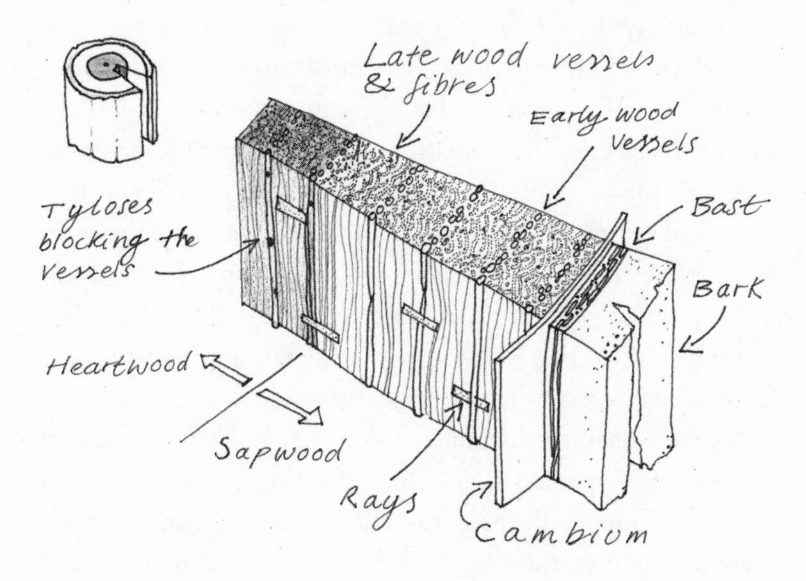


Elm wood magnified.

An interesting feature of elm is that the thickness of the sapwood band is extremely varied, even in the same tree and from one side of the tree to the other. Sometimes, there's a thin pale cream border of sapwood beneath the bark perhaps just three years' worth of annual rings, and sometimes the sapwood forms several inches of wood representing fifteen years of growth or more. In a highly irregular way, elm sapwood in some trees appears to be wandering amorphously in and out from the perimeter, crossing annual rings as it goes, reaching inwards in waves, shelves or even blobs. It seems to be completely random, though is a more common effect in mature trees. When dry the colour of sapwood usually becomes indistinguishable from the heartwood. The variation of thickness of elm sapwood even occurs to different profiles within the different limbs of the same tree. It all goes to show that it is best not to quote specific figures for elm trees, because you will very likely be proved wrong!

Traditionally, and often repeated in books, elm sapwood is considered as indistinguishable from heartwood in terms of timber use. That assumes that the elm tree was felled at the right time of year, when the tree was completely dormant and had the minimum of sappy nutrition in its cells. Trees that are felled at times of the year when the sap has risen are more susceptible to wood-boring insects and fungi. Traditionally for elm, felling would be in late autumn any time after Halloween, 31 October. Left too late in the winter, the sap would have already risen to feed the emerging blossom buds.

Milling right to the wane makes for very efficient use of timber as less is wasted, as well as providing wide boards that include the sapwood. There is a tradition of using elm for lapped waney-edged weatherboard on the outside of timber-framed buildings in many regions of England. The undulating lines

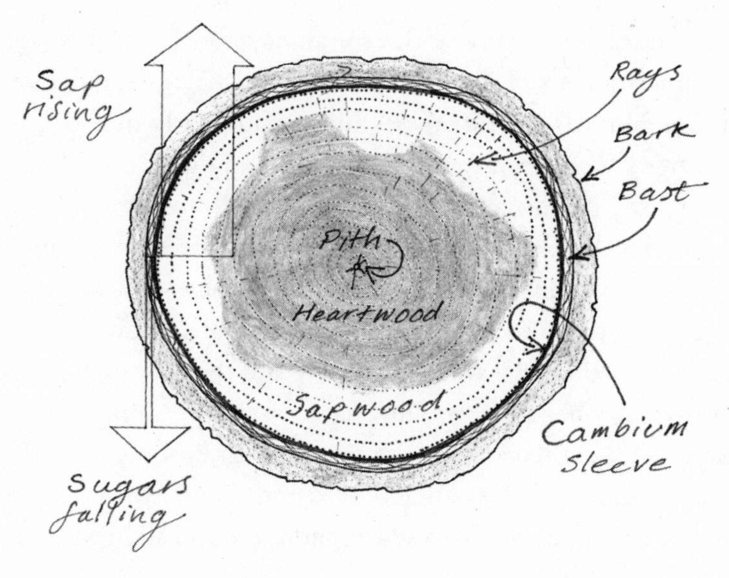


Cross-cut section of an elm log.

of the boards' edges have a striking visual effect. A cladding tradition like this only evolved because it was found to work. It takes advantage of the relative durability of elm sapwood. Used like this it can last for a century, which wouldn't be the case for other types of tree. Oak sapwood would rot much more quickly, for example, if used in the same way.

When a tree log is cut lengthwise, the vessels and fibres running along its length are exposed. Some of these will be sliced open to create a series of minute channels. This is known as the 'grain'. The combination of the species of tree, the speed of growth and the size of the vessels exposed decides if the grain is fine or coarse. The wood from the outer sections of an old tree will be finer-grained than from deeper inside, when the tree was younger and growing more vigorously.

A feature of the sloping spiralling alignment of the fibres is that the annual rings running up and down the tree may not be in the same direction as this spiralling grain. In other words, the lines that are visible may not be the same as the direction of the grain. The most cross-grained and gnarly timber is at the tree's base, even if the grain looks straight. You cannot always tell if a piece of wood is strongly spiralling or cross-grained by looking at the direction of the sliced annual rings. But you can tell if you split it because the split will always run along the direction of the grain. For example, when cleaving oak for making pegs, you sometimes find the split wood runs in a spiral, which you wouldn't have known by simply sawing along with a circular saw. The spiralling grain in elm can be seen on a log with the bark removed and allowed to dry. Small cracks will appear on the surface, and these will align with the spiralling grain.

In some elm tree types the pitch of the spiral varies and even its direction reverses. This produces interlocking grain. Some of the vessels will be sloping out of sawn wood and some sloping back in, which makes planing or chiselling 'with the grain' extremely difficult. The practical answer is that you have to reverse the direction that you are working, continually being aware of and responding to the grain. Alternatively, you can work diagonally, across as well as with the grain, in order for the blade not to snag or tear the surface up as you work.

The angle of the spiral and its alternating direction, clockwise or anti-clockwise along the log, leads not only to a wild confusion of fibres, but to bundling and thickening at certain places. Sometimes there are three annual rings per inch, sometimes twenty. It is part of the beauty of elm wood that the annual rings seem to grow at completely random thicknesses at different points along the trunk. This movement of the annual rings also makes the surface figure run wild. Sometimes the inconsistency is followed year on year

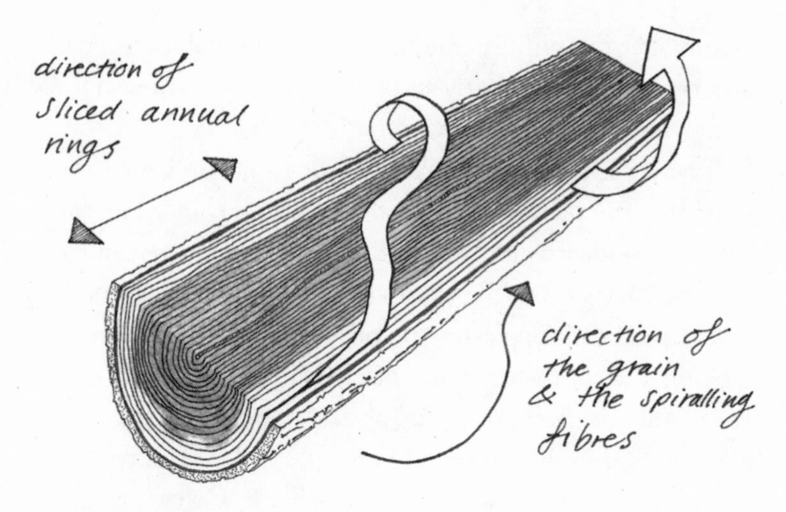


Wildly spiralling elm grain.

as the tree grows and produces the visual effect of a cresting wave or a ram's horn on the surface of a milled board.

The medullary rays radiating out from the centre are curtailed by the spiralling late wood fibres passing this way and that. So the rays of elm are thin, short and interrupted. There is no clear line from the outside of the tree to the inside, unlike oak for example, which has large, long rays. When trying to split a log, an axe rives open the line of rays towards the middle of a log and then follows the sweep of the grain down through the wood, paring the longitudinal fibres apart. If the rays are interrupted and there is no common line of grain fibre, the wood is fighting the axe, and the wood usually wins. This gives elm the reputation for being difficult to split. Fortunately, when elm is freshly felled and green, the cells are full of water and the cell walls are soft. They are no match for sharp metal tools and green, fresh elm mills, cuts and splits beautifully. It is only when dry that its tough, forbidding nature begins to appear. This is hugely significant as it makes dry elm the toughest timber available in the UK and means it is eminently suitable for pegged timber framed joints, as will be explained later.

That is the story on the inside of the cambium layer. On the outside a completely different thing is happening. Here new cells are produced that will transport the sugars generated in the leaves back down the tree to its roots and their surrounding mycorrhizal soil fungal system, or to wherever in the tree the growing cells of the cambium need energy. This layer of cells is known as the 'bast', or 'phloem'. These longitudinal cells are again a mixture of tubes and fibres. As the tree grows wider, these cells are flattened against the bark. Other groups of cells amongst the bast produce a corky substance. Every year the bast is renewed and the old bast and old cork cells become the dead bark. The older the tree, the thicker the bark. The bark is waterproof and gives protection from the cold, sun, animals, insects and microorganisms. Because the bark cannot grow, being dead, it cracks open into fissures as the circumference of the tree stem increases. Veteran trees have extraordinarily expressive bark that reflects the age and character of the tree that it enfolds.

Bast carries sugars, but elm bast contains protein as well. This gives it particular properties for weaving or twisting into a twine. This is more easily done in spring when the cambium is active and the bast peels away from it, leaving a wet, smooth surface. Elm bast rope has been used in Europe for at least nine thousand years. In John Evelyn's *Sylva: a discourse on forest trees,* published in 1664, he states that the bast of wych elms was made into a coarse rope for the navy. Elm bast was woven into textiles by Creek Indians in North America and was still being used in the UK in the early twentieth century, woven into mats for plant nurseries. Even more surprisingly, bast from some elm species has sufficient nutritional value to be dried, powdered and made into bread. This is a tradition across the globe, from China to Norway.

---

Dry elm wood is therefore hard, tough, elastic and strong. These properties have been well understood for thousands of years. That long tradition of using elm has now all but disappeared. One of the earliest uses was for archers' bows. One Celtic tribe in France was even named after elm, the Lemovices, or 'elm warriors'. This is the origin of the name for the Limoges region.

The once ubiquitous elm tree and its timber formed a part of everyday life for everyone, from its use for making water pipes to coffin boards. The objects that were made of elm, that our ancestors observed and handled every day, have been nearly completely lost from use. To understand the potential of elm in the future, it needs to be seen for its unique physical qualities. The following past uses are mostly redundant in terms of generating demand for the timber, but may expand our understanding of the potential for the timber and invite new applications.

The quality of toughness is why elm was used for hubs of wagon wheels and seats of chairs. Before the twentieth century, when tens of thousands of wagons and carts of all descriptions were in use, wagon wheels were made of wood, with an outer metal tyre. The outer rims were made of curved sections called 'felloes'. These were supported by spokes set at regular spacing around a wooden hub, or 'nave'. The spokes were drilled into the hub from all directions, irrespective of the line of the grain in the hub. Elm was used so that the hubs wouldn't split, no matter what direction a spoke was driven into them. Likewise, chair seats have chair legs and chair backs knocked hard into holes drilled near the edges. They must not split and loosen through the heavy handling they will receive. Again, elm is the perfect choice. Pulley blocks and deadeyes used for ships' rigging in the past also needed the same toughness of material. Elm can withstand impact shocks, which is why it was also used for threshing platforms in tithe barns as well as gunstocks.

Elm can also be milled very thin, and nailed or screwed into its end-grain, which would be a problem for other types of timber. Coupled with its varied figure, this made elm a popular wood for making boxes, chests, crates and of course, coffins. The tough interlocked grain and readily available wide boards of elm also made it suitable for lining the beds of carts, trucks and mining wagons that suffered constant knocks and dents during use. It doesn't crack or splinter easily. For the same reason, elm became a popular choice for floorboards. Flooring is one of the uses that could still apply in today's world, particularly used over underfloor heating pipes. Elm wears very well and generates wide boards when milled to the wane, so staircases of elm are another traditional use that could be continued.

The one drawback of elm is that it rots quickly in contact with topsoil or when kept damp in the presence of oxygen. A large variety of fungi occur on elm in the woods and fields. Elm fence posts will also soon snap off at ground level. Beams will rot through when built into damp masonry. Logs also seem to quickly draw fungus from the soil in a matter of months when left lying on the ground. If left too long sitting on the ground, the timber becomes doated with patches of almost white, soft rotten wood. Or, the core of a log may have become 'foxy' with areas of very dark orange, resulting from brown rot. Elm that has begun to rot also suffers from 'brashness'. This is when the timber snaps with very little effort. For this reason, elm needs to be milled as soon as possible after felling.

Fortunately, fungi will not grow on timber that is dried below about 18 per cent moisture content, so air-dried elm is safe from attack. Once milled and kept off the ground and well aired, such as in the frame of a barn, elm will be safe. As it dries it becomes lighter, tougher and stronger and will last indefinitely. Ironically, if kept underwater the same is also true. Thomas Laslett, writing in *Timber and Timber Trees,* published in 1875, states that if elm logs were to be stockpiled, they should be submerged underwater or buried in mud. Submerged logs are far less attractive to fungus as the oxygen they require for respiration is unavailable. This property of durability underwater had many uses in the past. Elm was used for pilings under bridges in London and on the sides of wharfs. It was used for water pipes in the thousands in major English cities in the seventeenth and eighteenth centuries. Cracks in pipes are obviously best avoided, so a hollowed elm log that doesn't split is an ideal choice. Elm was similarly used for water pump pipes and valves, as well as waterwheel rims and paddles. Lock gates, sluice gates and aqueducts were likewise made of elm wood. Naturally, the bottoms of boats would have timbers that were constantly submerged.

Just west of London, on an arm of the Grand Union Canal, boat builder Chris Collins has his boat yard. Chris has rebuilt narrowboats and barges at Troy Wharf for thirty years. He used oak planks for the sides and elm for the flat bottoms, as is the tradition. The *Albert,* originally built in 1926 and rebuilt by Chris in 1990, has 3-inch-thick elm boards spanning across its flat bottom. Some are 3 feet wide, which has the advantage of reducing the number of joints. One large elm tree provided all the timber he needed for *Albert.* He showed me a 3-inch elm bottom board he replaced from the original narrowboat. It had worn down to 1 inch on the gravels and silts lying on the bed of the canal. Other timbers would have splintered off, but the toughness of elm resisted this abrasion and splintering for sixty years.

The long timber at the base of a seagoing vessel, the keel, also had to hold numerous ribs and fixings without splitting. Elm is again a sensible choice. The keel of the famous British tea clipper, the *Cutty Sark,* was made of elm, but it was rock elm imported from Canada rather than English elm or wych elm that were previously more commonly used. Resistance to decay applies to timber on the inside of a boat as well, where there could have been several feet of water in the bilges of a large wooden ship flexing over the waves when sailing hard. Elm was therefore not just used for keels but sometimes for the first boards on either side of the keel, the garboards, as well. These planks would often need to be twisted to fit the shape of the hull. Another property of the elm fits this purpose. Like ash wood, elm can be easily bent into new shapes when steamed. Some pieces of elm are more bendable than others. In fact, when we were milling the elm for the Carley Barn, there would be the occasional piece that would have a remarkably stringy edge. The fibres were loose enough to appear as fur. These produced strips of wood that were strangely plastic, almost rubbery, and could be bent into shapes that would not spring back into their original form. I haven't encountered this cold-bending in any other timber.

Elm is traditionally thought of as a poor firewood, but not by those who dry it first. Like a lump of coal, it is difficult to ignite, but once placed on top of a fire it burns perfectly well. In fact, it stays steadily glowing like a big cherry-red brick for a long time. That makes it a good wood to burn once the fire is started and a great Yule log. The ashes of elm are high in potassium. That makes it ideal for spreading under apple trees, which need large amounts of potassium each year for producing fruit.

The quality of initial resistance to ignition is reflected in its use as an anvil block in a blacksmith's forge. Hector Cole, a master blacksmith in Wiltshire, related to me how elm is his timber of choice. Every wooden implement in his workshop was made of elm. When hot metal falls on elm it just smolders for a while, then the fire extinguishes. In the absence of an elm tree trunk, a fellow blacksmith once used a large oak log to support an anvil. A drip of metal landed on it and overnight the oak log caught fire and the anvil was on the floor in the morning. An elm anvil block just smolders and doesn't flame up in the same way. For the same reason, Hector beats out his metal bowls in a log of elm with hollows gouged out to different sizes. The first time hot metal is used, the elm scorches, but after that the burnt surfaces are stable. This resistance to combustion would likewise be the reason for the boards of bellows being made of elm.

Elm wood is one of the least acidic timbers. Comparing the sawdust of oak and elm with a simple pH tester kit indicates oak sawdust as acidic and elm as alkaline. Elm bark is also alkaline, whereas, again, oak bark is acidic. Perhaps the alkalinity gives elm a slightly disinfectant property in the same way as the alkalinity in lime wash. Elm also has a reputation for not tainting the food or liquids it comes into contact with. This may explain why elm was a suitable wood for chopping boards, butchers' blocks and food containers such as dough troughs. Or, perhaps, why it was used for wooden toilet seats. Similarly, the beautiful elm floor in a bedroom at Montacute House, Somerset, is supposed to have been selected because it withstood spillages of 'night soil'. There are usually good reasons for long-standing traditions, and I find it intriguing to find links with the physical and chemical properties of elm wood.

There were clearly many uses for elm timber in the past. In the early twentieth century, Henry John Elwes travelled the length and breadth of the country with his co-author, Augustine Henry, cataloguing trees and their uses. This took many years, wearing out six motorcars as they did so, and led to the publication of the seven-volume work *The Trees of Great Britain and Ireland* between 1906 and 1913. They recorded some extraordinary trees, including elms. Some were 150 feet tall. That would be 50 feet taller than any of the trees growing in the woods where I live. He was also able to interview the people who had used elm firsthand. This was a very interesting period to be writing because it predated the First World War, at a time when traditional country craftsmen were still able to talk about the vast range of products mentioned above. Elwes and Henry's travels also predated the arrival of the first wave of Dutch elm disease that removed so many trees from the landscape.

This period is another fulcrum moment in the story of rural crafts, as it represents a time when village carpenters were still finding their timber from the countryside around them, and still had a high degree of knowledge of different timbers and a proficiency with using hand tools. Many, like chair 'bodgers' working in the beech woods of the Chiltern Hills to the west and north of London, would be comfortable working outside as well. This all changed with the accelerated industrialisation and mechanisation that followed the outbreak of World War I, as well as the death of a generation of craftspeople in Europe. This terrifying destruction began just one year after the publication of the final volume of *The Trees of Great Britain and Ireland.* Elwes writes: 'Most country wheelwrights who have a reputation for the durability and soundness of their wagons prefer to select the trees growing in their own district.' At the time of publication, not only was being a wheelwright a common occupation, but the timber they used was local. A carpenter had to get to know the trees of their locality and base their knowledge on that intimacy.

The arrival of readily available imported timber made that knowledge redundant. Yet, the change did not happen overnight. From the turn of the nineteenth century onwards, elm was increasingly imported from Europe and North America. Good-quality cheap imports suppressed the price of home-grown elm, making it less financially attractive to look after the quality and maintain the stock of local trees. The earlier enthusiasm for planting elm for profit, shown by Georgian writers, seems to have waned by the time Queen Victoria was crowned. The resources from around the world were also seemingly endless. In 1875 Thomas Laslett estimated that up to 250,000 cubic feet of rock elm were arriving in Britain from Canada every year. Rock elm, along with American elm, continued to be a competitor to local supplies well into the twentieth century.

By the 1970s, when Dutch elm disease arrived, the vast majority of traditional uses for elm, like cartwheel hubs, had become tiny niche markets requiring very small volumes of timber. There were a few exceptions: bellows still were being made with elm; solid elm coffins were still in demand; and the famous furniture manufactured by Ercol continued to be produced. The waving grain of elm makes it a fantastic wood to sculpt, with a beautiful example created earlier in the century by Barbara Hepworth, which can be seen at the Ashmolean Museum in Oxford.

When certain elms are infected by Dutch elm disease, the timber will have a dark brown edge to the annual ring just under its bark that year. Large elm trees that I have milled sometimes show this dark annual ring in their timber from thirty or so years before, and might suggest that these trees had survived exposure to the disease several decades previously. It is conjectural, but the evidence of tolerance of Dutch elm disease may be recorded in the timber of the few surviving elms.

Speaking to timber merchants and sawyers who were felling and using elm at the time of the arrival of Dutch elm disease, you get the sense of a sweeping wave of destruction across areas where elms were the predominant tree cover. Places like Frome in Somerset. Andy Coombes was involved in felling and marketing elm there at the time. He describes the disease simply as 'traumatic'. The crown of a 100-foot-tall elm would be slightly brown on Monday and the whole tree would be dead by Saturday. He remembers selling the best elm to Germany for veneers and the bulk of the rest being sold as 6″ × 6″ 'chocks' to act as lintels for the pit props in the coal mines of Nottinghamshire and Yorkshire. The price of elm collapsed as the market was flooded in the early 1970s, to be followed by a collapse of supply as the vast majority of the large elms had gone by the mid-1980s. Whole woods of elm trees were removed from the landscape, sometimes just piled up and burnt.

Bob Wilden, a retired Suffolk saw miller, felled and milled so many elms in that county at the time that he must have the best firsthand knowledge of elms in England. He remembered getting a call from a lady in Ickworth near Bury St Edmunds. She was in her nineties and said there was an elm in her garden overhanging her house. She was worried about it, so he drove over to see if he could reassure her. In the end Bob felled an English elm tree that was 10 feet across at the base and provided him with 1000 cubic feet, or 25 tons, of sound timber. It must have been one of the largest deciduous trees in the country at the time.

Why there is a localised tradition of elm timber-framed buildings in the East of England is a bit of a mystery. For a timber to be useful for framing it needs to be strong, tough and durable. The first two of these three is true of elm, particularly for forming tight joints that will not split apart, but the durability is a drawback. Oak is the traditional and most common timber-framing material in Britain. Oak is certainly more durable. It may be that oak trees became scarce in these Eastern landscapes but there is no evidence for that in the countryside today. Perhaps they simply preferred elm. Perhaps they discovered how easily and quickly it grew to a useful size and that it produced knot-free timber that is ideal for spanning between the supports of a timber frame. There are certainly many eighteenth century writers extoling the virtues of elm as a useful and profitable timber tree. As long as the roof of an elm building is kept in good repair, the timber will not rot. But during the agricultural depression years of the late nineteenth and early twentieth centuries, many barn roofs suffered from poor maintenance, and roof leaks meant that some elm timbers succumbed to rot. The Great Barn at Wallington of 1786 is an exception. Whatever the reason, there are enough examples of elm barns to establish elm as a branch of English timber framing history.

Different types of elm tree produce very different types of elm wood. It is difficult to get a reliably good impression of what was used for different historic purposes, if any distinction was made at all. Fortunately, two writers took the trouble to put their memories to paper and give authentic accounts of craftsmen using local elm. Walter Rose wrote about his carpentry workshop in Buckinghamshire during the late Victorian period in *The Village Carpenter,* and George Sturt in *The Wheelwright's Shop* wrote about his business in Surrey during the same period. Walter Rose describes only three types of elm tree: English elm, wych elm and Dutch elm. George Sturt describes just one: English elm. Given the large number of different hybrids in the open countryside today, including substantial trees, it seems unlikely that they did not exist or were not used 150 years ago. It is curious that they are not specifically mentioned.

Although today there is a distinct elm type called Dutch elm, it is very odd that it does not actually grow in the Netherlands, which would be expected if the tree originated there. However, in the English language the word 'Dutch' can also mean unintelligible or strange, as in 'double Dutch'. It seems possible to me that the term Dutch elm was historically given by rural people to elms that were simply neither wych elms nor English elms, that is, wildings that had been the result of a random cross-pollination. In his book *Forestry and Woodland Life,* published in 1947, Herbert L. Edlin wrote about wych, field, English and Cornish elm, and, 'a number of hybrids and intermediate types that are described collectively as Dutch elm.'

Since then, the distinction between the numerous elm tree types has strengthened in terms of botanical taxonomy, but this has not necessarily translated into trade names for their different timbers. Confusingly, the names of living trees don't always transfer to the names given to their sawn timber. Even today lumber is often described in generic terms. In England oak is 'oak', even though there are two species of oak tree growing here.

Thomas Laslett wrote in the late Victorian era of only three sorts of available native elm timber: red elm, white elm and sand elm being sawn from English elm; wych elm, and Dutch elm, respectively. The *Encyclopaedia Britannica Eleventh Edition,* published in 1911, also linked Dutch elm with sand elm, referring to a 'Dutch elm or Sand elm'. Perhaps there were just three categories of elm in the timber trade: the red elm, the white and the sand.

Red elm from English elm trees is no longer available because all but a handful of large trees have died of Dutch elm disease. But it was one of the most numerous timbers in the past and seemed readily available even as the economy and population grew. English elm is fast-growing with a coarse grain, a deep brown colour and a varied, swirling figuring. It also potentially grew into straighter stems and longer lengths than wych elm, making it more suitable for water pipes, construction and for flooring. Wych elm tends to grow with a forked trunk, providing timber for uses where long lengths aren't needed. White elm, from wych elm trees, usually has a finer, straighter grain and slightly lighter colour than red elm, and is an easier wood to work with. It is stronger and heavier than red elm and seems to have been preferred for chair seats and for steam-bending into different shapes in boat building. White elm is still occasionally available from areas of Scotland, where wych elm trees appear to have a degree of tolerance or field resistance to Dutch elm disease. Both types of elm seem to have been used to make wheelwright's hubs and the whole range of general uses.

The third type of elm timber, sand elm, is more complicated. Thomas Laslett described sand elm as inferior to the other two, being subject to star shake and less sought after. In contrast, Walter Rose rated 'Dutch' elm wood as a 'kind-working, mellow wood … all woodworkers incline favourably towards.' They must have been writing about different trees with different quality of timber. Or, as I suspect, if Dutch elm here refers to a range of different hybrids, then possibly both accounts could have been correct.

Unlike white and red elm, sand elm timber would have had very variable characteristics, depending on what hybrid or how 'Dutch' the elm was that grew in the hedges and woods. These hybrids are often extremely localised, so could also have enjoyed a local reputation, complimentary or otherwise. Unfortunately, this local knowledge of the quality of different timbers from different hybrid elms has been lost. After the transformations of World Wars I and II, twentieth-century writers continued to list uses of elm that were already historic, if not obsolete at the time of listing them, so they would not have been able to interrogate the craftspeople who had actually felled and used the wood. The particular variety of elm, its source, its local name and its preferred qualities for a particular product remain unknown. There may have been strong local traditions of use for the different elm types, but those traditions are gone. The dough trough or water pipe had long since passed out of the workshop door.

The red and white elm, as well as rock and American elm that together formed the bulk of historic elm timber use have all gone, and their timber is unavailable today due to Dutch elm disease. It is a strange twist of fate that the only remaining large elm trees in Southern England are the many 'Dutch' hybrids that generated the timber we once knew collectively as sand elm. The different qualities of the timber of the many hybrids growing into usefully sized trees today will need to be rediscovered and worked out from scratch. People using elm sourced locally, as we did for the Carley Barn, really are pioneers, rediscovering what the timber of their local trees looks like and the subtleties of what it can be successfully used for.

SEVEN

# Boxed Hearts and Waney Edges

It is hard to say why I find the milling of tree trunks so compelling. Certainly it's the sight of the beautiful heartwood within each tree, but it is more than that. More than the intoxicating sweet honey smell of some of the freshly milled elm. Milling a tree is a turning point. In this story it transforms our lead characters, the elm trees, into a new and completely different state. Milling is the decisive act that cuts the log into the particular shapes that had been imagined whilst first walking in the wood. Too many mistakes and there will be an excess of one dimension and a dire shortage of another. Too much taken off a particular stem and the beauty of the undulating wane on the edge of the tree is lost forever.

For the volunteers of Barn Club who experienced the wood where the trees grew, it also felt like we had an unspoken contract to do right by the trees and bring their essence into the things that we were to make out of their timber. At the moment of milling, something new appears that you form a strong relationship with, as you handle it and cut the various carpentry joints. As you return to a particular piece of wood, you might also remember the wood and even which tree it came from.

The mobile band saw mill was hired in from Essex, along with the miller, Graeme Smith. Graeme has been milling for thirty years. His mill, a Wood-Mizer LT40, has served him almost as long as that. Graeme is part mechanic, part doctor, attending constantly to its needs. It's a well-looked-after machine, though much of the orange paint has long since worn away. It has had an updated version of its engine installed since the milling at Churchfield Farm, but at the time it started up with a great cloud of smoke and diesel fumes. It didn't take much fuel to run, but the fumes were enough to make you want to stand on the upwind side of the engine.

Lengthy discussions were had to position the mill so that the bed was at the right inclination, the stacks were aligned, but, most importantly, so that the sawdust would not be blowing back into Graeme's face. Graeme doesn't enjoy sawdust blowing in his face. His focused attention and pugnacious mill-side humour were exactly what were needed as we steadily progressed through the stacks of logs in the farmyard. His ear defenders had built-in speakers playing music that only he could hear, so everyone became used to observing Graeme's sidesteps and shuffles as he walked up and down following the mill's blade and doing an occasional little dance. Milling is noisy work and takes considerable concentration and organisation. Fortunately, Bruce Carley and ample volunteers were available and keen to be involved and witness the transformation of logs into lumber. Apart from the miller, the operation needed two people to roll the logs, two to carry the timber away, and ideally two more to sweep off the sawdust from the freshly sawn wood and stack it in the right places near to the framing yard or under the two huge lime trees at the back of the tea shed.

We were using a diesel-powered mill, but in previous centuries, the logs would have been hewn into shape with axes or sawn with long, two-man ripsaws. The slightly undulating scalloped surfaces of historic beams are a sign of them having been hewn with an axe. Hewing with axes is fast, powerful work and gets the job done and over with. Side axes are used, which have a bevel on only one side. With practice they can be very accurate.

For the Great Barn of Wallington of 1786, the elm was first hewn square using side axes. This produced the main posts and tie beams. Then some of these squared baulks, or cants, would have been levered and rolled above a pit to be ripsawn into the various smaller timbers. One man stood on top of the timber being sawn and one man underneath. One standing in the fresh air and the other below, standing in the sawdust. Sawyers travelled from village to village sawing the timber that the carpenters had procured from their local woods. Sometimes the pits were no more than staging set on the sides of hills, down onto which the logs were rolled. This was common in the West Country. The tradition for the sawing to be done on-site or in the carpenter's yard continued briefly into the industrial age, when travelling steam engines took over from the pit sawyers. These were soon superseded by stationary mills, and by the whole system of timber merchants providing timber to be delivered direct to a workshop, which persists to this day.

To produce the timbers we needed for the Carley Barn, we wanted to follow the tradition of sawing on-site. We could have set up a pit and hewn and sawn by hand, but it would have been a big task. It would also have been very hard, strenuous work. There is a balance to be struck between mechanical power and quality of experience in each process. I made the decision to use powered tools to do the milling for this project, whilst still handling the logs and timber by hand rather than with a telehandler or forklift. We would then switch to hand tools for the less physically exhausting and more intricately skilled task of cutting the joints and raising the frame. We also had a constraint of doing everything within about six months of reliably dry weather. So, it was decided to use a combination of chainsaw mill and mobile band-saw mill to keep to our programme of completing the carpentry before the autumn.

It is surprisingly easy to roll logs along the ground, but not by trying to push with your arms. The tool for this job is what makes the difference. It's called a cant hook. It is a very simple device, and with the help of its leverage, even very large logs can be rolled onto the lifting arms of the saw. This was made even easier by having tramways of 8-inch poles in front of the mill for the logs to roll along. These extended under each stack of elm logs, keeping them out of contact with the soil and reducing the opportunity for fungus to take hold.

There is a large range of thicknesses and sizes of timber in a traditional barn that makes the task of planning ahead relatively easy. If the list consisted of only one or two sizes, say 8″ × 2″ and 6″ × 4″, there would be much more waste. It always impresses me how little is left over from milling the trees for a traditional frame. So little goes to waste. It's a combination of the sizes of trees that were felled – many small, a few medium-sized and very few large ones and the handy range of timber sizes to be cut from each. Even the slabs removed from the outsides of the logs, the slab rails, can be re-sawn either into cladding boards or studs.

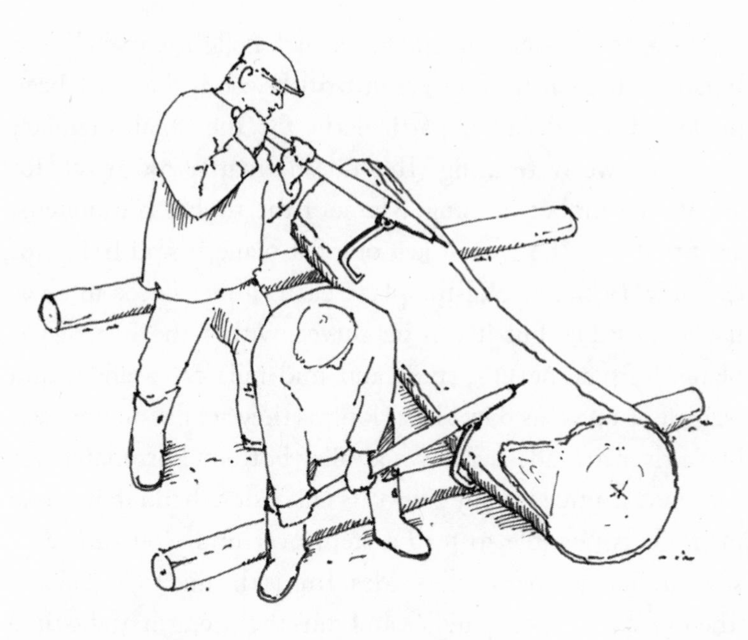

Rolling logs effortlessly with cant hooks.

There is a great sense of anticipation at the beginning of a milling session. An early start, warm cups of tea and a degree of nervous joking occurred as we gathered in the tea shed for the beginning of a week's work. Gordon Gray, an antiques valuer, a great team player and someone with a passion for learning about traditional crafts, recalled the day we started milling:

> *A cold but a clear blue, cloudless sky and the stage was set with Tewin church as a backdrop and an ancient cedar reflected in the big pond. We were in for a great day. It is such a powerful thing to watch. The speed and ease with which your log is cut in half is mind-boggling. The machine can then roll the log on the bench to position it for the next cut. In due course a willing team of hands took away the tie beams, all cut to size ready for marking and morticing.*

Most of the stems in the stack had a slight curve. This is typical of the hardwood trees in British woods that have been neglected for decades, particularly for the small-diameter trees that we were using. This doesn't matter too much for traditional timber framing. Many of the timber components are fitted into the face of just one flat plane, a wall frame for example. In the face of the plane each timber needs to be as flat as possible, but it can be curved within the line of the plane. Rafters, purlins, studs and mid-rails can wander and wriggle as much as they like as long as they are flat on one face, like a river meandering down a valley bottom. Fortunately for carpenters, a tree usually grows its curve or S-bend in just one plane. It is possible to turn a stem over until you can see a straight line down one of its sides. This is the line to mill along. The fibres will be cut parallel, and thus the strength of the timber is not diminished. Wriggly rafters and even floor joists are therefore very strong. By using slight curves where you can, the amount of useful timber from deciduous woodland thinnings is approximately doubled. That, in itself, is a substantial reason to promote traditional approaches to structural carpentry.

When a living tree is first felled its vessels are full of water and the timber is termed 'green'. After milling, the water in this green wood slowly evaporates. Release of moisture from the wood is called seasoning. Some timber can have over 50 per cent moisture content when felled. This is the weight of the timber sample less its weight when it has been oven dried, divided by the weight of the same oven-dried sample. In extreme cases, where the cells of the timber are completely filled with water, there is more weight of water than wood itself and the figure goes over 100 per cent. The elm I encounter in Hertfordshire is about 40 per cent. This quickly dries off the surfaces, but can take years to dry from deep inside a thick beam.

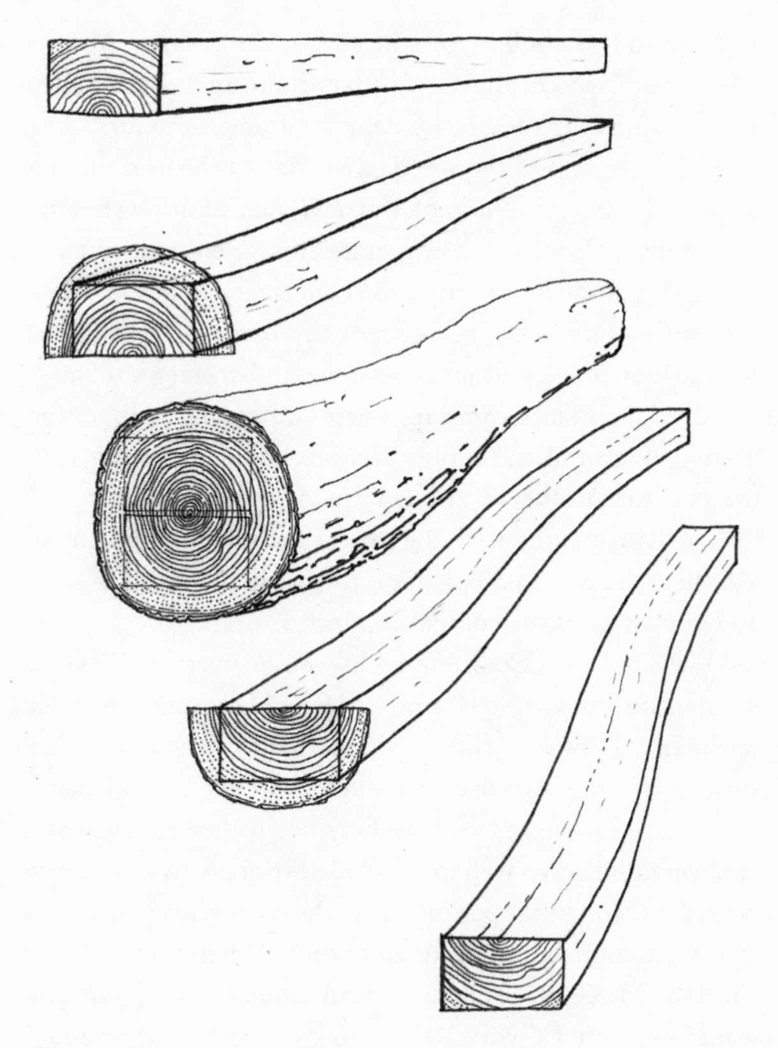

A curved log can be milled to give one flat surface plane.

There are two stages in drying, which is very significant to a timber framer. Surprisingly, we want the timber to stay green until the building is raised. In fresh green timber the cell water, known as free water, gradually evaporates until the moisture content is about 30 per cent. At this stage the cells and the milled timber are still all the same shape. Whilst the timber is green, above 30 per cent moisture content, it is stable, soft to cut and ideal for timber-frame carpentry.

When the cells of the wood have dried out to below about 30 per cent moisture content, the next stage of drying begins. The water molecules that evaporate are those that are bound through molecular attraction to cellulose in the wood fibres, termed 'bound water'. This second stage of drying is when the cells begin to buckle and therefore the timber starts to change shape. The surfaces can cup, warp and crack. Evaporation keeps going until the timber has reached equilibrium with the moisture in the air and is termed 'air-dry' or 'seasoned'. This obviously varies with the season and the humidity in the air where the wood is kept, but is usually in the range of 13 to 16 per cent moisture content in Hertfordshire.

With its interlocked grain, elm seasons very well and is resistant to cracking and splitting. Its wide boards are stable when dry. However, the same interlocked grain pulls the surfaces out of shape as it dries below 30 per cent, so floorboards need to be stacked carefully when they are seasoned and weighted down well to avoid distortion. A rule of thumb is that it takes a year per inch for a heavy timber like elm to air-dry, though it is much quicker for less dense timbers like western red cedar. It is interesting to note that timber is safe from fungal attack when its moisture content is below about 18 per cent, so air-dried timbers that are kept dry don't need fungal preservatives.

The sawn timber also shrinks in its outer dimensions as it dries below 30 per cent moisture content, but not by the same amount in all directions. It shrinks the least along its length. So little, perhaps just 0.1 per cent, that it can be ignored. Wood shrinks most in the direction of the annual growth rings (tangentially) and for most species of hardwood about half that amount across the annual growth rings (radially). For the carpenter, what this means is that a rectangular sectioned timber might have perfect 90-degree corners when it is first milled, but will distort into a trapezoid as it becomes air-dry. With elm there is less difference between radial and longitudinal shrinkage. This is partly why elm is dimensionally stable whilst drying, though there will be exceptions. There is a useful rule of thumb that as timber dries it straightens the curve of the annual rings as seen in a cross section. So, a board cut near to the surface bark, a crown-cut board, will cup as if it wants to straighten its annual rings. Conversely, a board cut exactly along the radial line, or quarter sawn, will be the most stable.

If the timber is allowed to dry slowly, there will be the same overall shrinkage, but fewer cracks and distortions. So it's good to stack timber out of direct heat and sunshine. The shade of trees is ideal. More water evaporates from the end of the timber through the end grain, than through the sides. To prevent the cracks, or 'shakes', it is a sensible idea to seal the ends with some PVA glue slightly diluted with water. Wood cracks in a number of ways, and none of them particularly helpful to carpenters. Heart shake, star shake, cup shake and radial (perimeter) shake are the names given to the shape of the cracks appearing in the end of a timber. If the bark is taken off a log, the outside starts to dry more quickly than the inside, and radial shakes appear starting from the perimeter edges and working their way towards the middle. These radial shakes also show up as longitudinal cracks along the surface of sawn timber following the line of the fibres, particularly where the outside surfaces dry more quickly than the middle. In a large-section timber like a beam this is almost unavoidable. If the log has been milled to follow the grain, then both heart and radial shakes are in alignment with the shape of the timber and little strength is lost. However, if a strongly curving piece of wood is cut to a straight line, the cracks will appear diagonally on the finished timber's surface, and in extreme cases the timber will snap. It's a very good reason to follow the line of the tree, not a straight edge!

Sometimes large trees can also develop cup and star shakes as they grow, even before they are milled and start to dry. Sweet chestnut is renowned for cup shake, and I have found star shake in some large specimens of sand elm. These types of cracks tend to extend as continuous splits along a log and are troublesome as the timber just falls apart as it is lifted off the mill bed.

Since drying cracks only appear when all the free water has gone, at about 30 per cent moisture content, there is a period of about 3 to 5 months after milling whilst it is still green to use the wood for framing. The alternative is to use only air-dried wood, that is likewise dimensionally stable, but much tougher to cut and chisel and would require waiting for at least ten years for the timber of a large beam to fully dry and become stable.

That alone is a very good reason to use green timber. However, it is also very satisfying to experience the immediate continuity of tree-to-mill-to-frame. A fellow timber framer, Rick Hartwell, described his experience of this to me. On a large project in Sussex the frame was going up well, but towards the end of the raising they discovered that there was one floor beam missing. Luckily the customer owned some woodland out at the back and Rick set out to find a suitable oak tree. On that same day the tree was felled and milled with a chain saw mill. On the next day the joints were

Different types of timber shakes.

marked and cut. On the third day the beam was in place, locked into position amongst its fellow pieces of tree. It was one of the best framing experiences that Rick can remember.

Working with freshly felled timber also gives the greatest length of time before surfaces of the joints begin to distort as they dry. The last thing you want is to find that the joints no longer fit tightly together. Once the frame is up and all pegs knocked firmly home, the worry is over. The forces bearing onto the shoulders hold the surfaces flat, and it doesn't matter if the joints lock tight. Green wood is also softer and much more gentle on the tools and wrists. It cuts like cheese. Elm, especially, needs to be cut green because the tangled interlocking fibres make it so tough when dry. A handsaw can easily get stuck, even when trying to cut the end off a dry cladding board. The elm seems to close in and grip the saw blade like a vice.

Most timber will crack around its pith as it dries and generate heart shake. Therefore, small sections that are milled very close to the centre will break up as they dry. Sections cut from near the edge of a tree will be more stable and free of shakes. As it happens, elm is less prone to cracking than oak, and the sand elm used for the Carley Barn proved stable across its surfaces as well. Though, it does depend on how wild the grain is. If the grain is wild, it will distort more as it dries, producing a slightly undulating surface. Although some pieces of wood twist like a corkscrew as they dry, they rarely change shape along their length. Some might deflect into a bend if they are overloaded whilst still soft and green, but as the timbers dry they stiffen, and the risk of deflections is reduced. Other curves in the timbers of an old barn are there from the moment they are milled. As we have seen, these might reflect the bent shape of the tree itself. Or, curves may result from forces in the wood being released as the mill blade passes through.

Trees grow with a slight tension in each annual ring of fibres, rather like the guy ropes of a tent. When the tree is milled, the balancing tensions are released. If a deciduous tree is cut exactly down its middle, the two halves will spring away like two bananas. To avoid this springing, the tensions have somehow to be retained in the finished timber. This can be done by boxing the heart. So, for a small 12-inch-diameter tree, 4 sides are trimmed away, leaving the middle, the heart, intact. This sized tree can produce a boxed heart beam 9 inches deep by 9 inches wide. Boxing the heart makes sure that the timber stays straight, which might be needed for example to support a flat floor. If the frame is to include a lot of boxed heart timber, then the different-sized components can be matched with different-sized trees. It means that small, young elm trees can be used right down to 5-inch diameter. These are used to produce 4-inch-deep boxed heart rafters or floor joists.

So, with the milling list to hand, and armed with the knowledge of all the different ways that the timber can shrink, shake and bow after it is sawn, the milling proceeded like a well-choreographed dance. Each log was rolled to the arms of the mill and effortlessly lifted into position with the

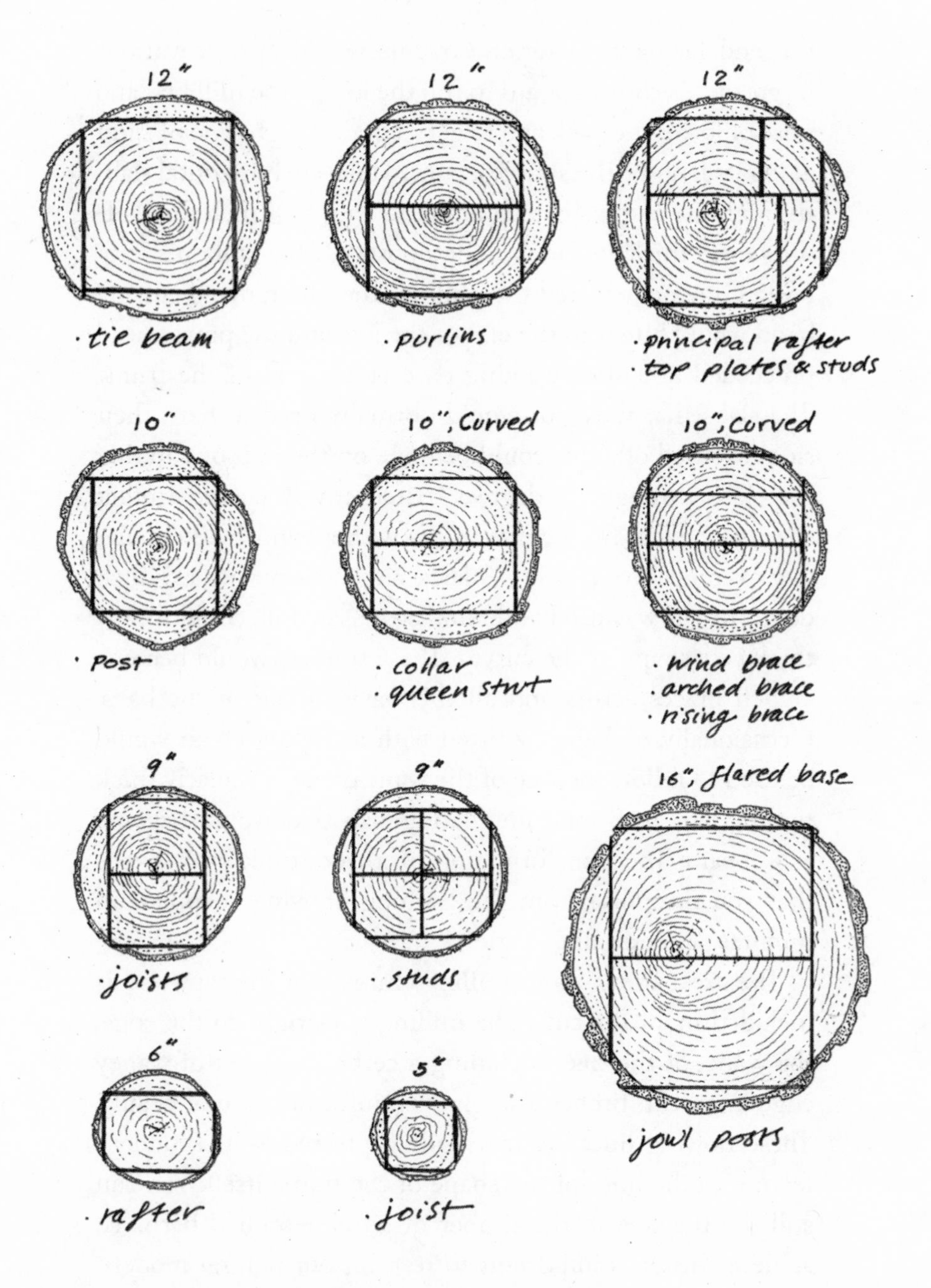


Different-sized timbers from different-sized logs.

top end facing the sawyer. Graeme would operate various levers on his control board to roll the log on the mill bed and raise or lower the back end to get a good, straight line of sight along its length. The simplest cuts to take are horizontal slabs through the trunk. These could be as thin as ⅝ inch for cladding or as thick as 7 inches for a post. With a pull on a long lever, Graeme powered the mill up. The fast-moving cutting band would bite into the end of the log, and the power head proceeded at a slow walking pace to the end of the trunk. The slabs that were cut would normally need to have their sides squared off. This could be done on the mill by rotating the slabs through 90 degrees. Alternatively curved timber shapes were cut from curved slabs with a hand-held circular saw over a pair of trestles. A large circular saw with a 5-inch depth of cut was used and, unlike the sawmill, could follow the lovely shape of the curves. These timbers would become arched braces, struts and all the wavy timber of the barn. Occasionally, a chainsaw fitted with a ripping chain would be used to follow the line of the wane on a particularly thick timber, like an 8-inch undulating slab that would become a tie beam. The transformation of a tree trunk into shapes that you recognise from the framing drawings happens in minutes. It seems like magic.

The great thing about milling elm is that the sapwood is included in all the cuts. The milling goes right to the edge, the wane, of the tree. Including a certain amount of waney edge in an elm timber also gives an informal natural beauty. These flowing lines are randomly expressed as a secondary feature to the lines of the shape of the frame itself. You can still see the tree in the timber of an elm-framed barn. To achieve this, it is important to restrain our natural modern tendency towards perfection. Moving the adjustment of the blade just half an inch on the saw mill can make the wane disappear and produce a completely square, uniform timber, which is exactly what modern sawmills aim to produce. It's nice and perfect, but the character is lost, as well as that extra half an inch of usable timber.

With other trees, like oak, the non-durable sap is generally avoided along with the pith, and so the total useful volume of timber from a 14-inch-diameter oak thinning is literally half that of an elm of exactly the same size. Not only that, but larger dimensions of timber can be sawn from an elm than an oak of comparable age. All the timbers for the Carley Barn could have been cut from trees of 14-inch diameter or less. An elm structural frame can therefore be harvested from trees of a much younger age, in theory as young as twenty-five years compared to the seventy years of conventional modern forestry.

To move heavy timber around by hand is hard work. Walking whilst carrying a load is bad for our backs and should be avoided in almost every circumstance. So, instead, the sawn timber is loaded onto a timber trolley and wheeled to where it has to go. A timber trolley can easily be made out of a small car trailer. A new deck has to be constructed higher than the level of the bed of the trailer, so that the timbers travel at about waist height. A batten fixed to the front and back of the deck holds the timbers off the bed enough to safely get your fingers under them to manoeuvre on and off. Finally, a front leg is added so that the trolley doesn't fall forward when loaded.

As the number of sawn pieces mounted in the yard at Churchfield Farm, they were ticked off the milling list and arranged in stacks according to their place in the different frames. These stacks were placed around the framing yard in front of the barn itself. The front and back wall frames were placed in two separate stacks, each gable frame and each of the intervening cross frames in others. And so on. Some, like the purlins and rafters would not be needed for a while and were stacked under cover of the lime trees where it would be cooler, and slow down the speed of water evaporation to ensure that the wood would stay green for as long as possible. Each layer of timbers had 2″ × 1″ battens underneath, and all the stacks were covered with sheets of plywood to keep the sun and rain off.

As the log pile grew smaller and the timber stacks grew larger, it became apparent that we were one tree short. One of the long stems earmarked as a tie beam proved to be problematic. This was the second of the logs that had a surprise hollow embedded deep inside the tree up at its top end. The tie beam had been milled without a problem or indication that a bird's nest was still inside. It was only when the excess length at the end of the beam was sawn away that the hollow interior with a hidden bird's nest was revealed. It was an impressive find, but not one we wanted. There was no spare length in the beam, so it was a question of including a bird's nest in the beam or finding a new piece of wood. After a lot of head scratching, no way could be seen to re-saw the log to position the nest cavity where it would not weaken a joint. We let the matter rest for a while to see what might turn up. And an elm tree did just that.

Later in the spring, whilst walking along a footpath near home, I noticed an uprooted tree across in some woodland. It had a disposition of branches that looked familiar. It took just a quick close inspection to see that it was an unusually tall and skinny wych elm. Luckily, it measured just big enough for a tie beam. It was growing in Cooks Wood, a few fields away from Churchfield Farm. Through asking around, I discovered that the owner lived nearby. He was surprised to hear that it was elm and was happy to let me have it in return for removing the problem of a hung-up tree. However, to safely ground the tree and mill it with the chainsaw in situ would require some deft work.

So, one sunny spring day Bruce, Gordon and I, who all have chainsaw experience, drove down the curving valley with our equipment in a trailer to the edge of the Cooks Wood. Mick, the owner of the neighbouring farm, was happy to let us drive across his meadow stocked with long-horn cattle and offered to lend a hand and his cable winch if necessary. He was amused by our unusual mission and promised to call by later on to see how we were getting on. I am glad he did, because a hung-up tree is a tricky puzzle. As Gordon noted:

> *Our chosen tree was well placed no more than 50 yards into the wood. We had a sack trolley with us so once felled we should have little difficulty in getting the milled beam to the trailer. It had recently blown over, as sometimes happens, and it was firmly enmeshed in the canopy of its neighbours at an angle of about 60 degrees and was still partly rooted. I began to understand Mick's parting quip: 'Watch out for the widow-maker!' Kick-back, roll and sudden stem failure can catch out even the best woodsman.*

But the day was beautiful, the birds were singing and we felt in no hurry. Just as well. The tree seemed very happy to have its branches enmeshed with its neighbour, and every foot of distance that we moved the tree's base managed to lock the crown of the elm deeper into a tangle of branches. Eventually, with the help of Mick's winch with its pulling force of several tons, the tree came down with a sudden crack and a crash. Because of the time of year the tree was heavy with sap and managed to split along at least a quarter of its length. Our luck held, though, and there was just enough length remaining in the stem to get the beam that we wanted.

Milling in situ is great. A wheelbarrow is all you need to bring the chainsaw mill to the tree. The setup is simple. First, a guide plank is fixed to the topside of a felled tree and then, with an aluminum frame bolted to the guide bar of the chainsaw, the top slab of wood is sawn off. This leaves a flat, straight surface for the rails of the mill frame to glide over for the next cut with the saw set at the required distance beneath it. We milled a tie beam in a couple of hours. With all the excess weight cut away, two people could just about lift one end at a time so that the beam could be balanced over the wheels of a sack trolley. Wheeling it out of the wood was made easy by pulling on a rope attached to the trolley's wheel axle. Very little damage was done to the ground cover, and all the sawdust and offcuts were left in the wood where they would rot.

Considering that a tie beam is the heaviest timber in a barn, it seems to be very little trouble to cut and move. Compared to the machinery, transport and fossil fuels involved in obtaining an elm tie beam by modern means, our collection of the beam from a wood nearby seems worlds apart. At the edge of the wood the beam was manhandled onto the trailer and driven the mile or so to the site. Our one regret was that we didn't arrange for the whole Barn Club team to haul the timber up the meadow and walk the beam gently along the back lane to Churchfield Farm ourselves: doing the whole thing from woodland source to farmyard site by hand.

At the end of the milling week with Graeme we all felt knackered, but very pleased with what had been achieved.

All of the primary frame timbers were accounted for. There were a few outstanding studs and rafters to reckon with, but this was expected. Generally, it is good to mill in two shifts, a few weeks apart. It is difficult to gauge how many of the smaller secondary timbers, the studs and the rafters, will be produced as by-products of milling the primary frame. The answer is to see what you can achieve and make up the difference later. If there are any timbers yet to be ticked from the milling list they can be sourced and shortfalls in numbers made up, again felled during the winter months. Secondary timbers are much smaller in section than the main framing timbers and consequently the trees that produce them are much lighter. It is possible to fell and extract them by hand, then haul them away in a car trailer. There are also local woods of elm trees growing tightly together that would benefit from thinning, so finding local sources is not too difficult in the East of England.

On a farm in Cambridgeshire near Baldock, there are hedges, hedge rows and small woods filled with semi-mature field elms. Guy Slater, the owner, states wryly that his soil will grow elms but not oaks, and, indeed, his farm is covered in them. His ancient barns have elm timber frames as well, so this dominance of elm trees could have been the case for generations. Guy was delighted to sell the thinnings to be made into something useful rather than as firewood. He showed us Boys Bridge Wood, which borders a tributary of the river Cam and that was where we found our trees.

They were all grown from coppice stools, so as at Bushy Leys Spring, the strong root system would power the next generation of regrowth without any need to replant. I felled sixteen trees that looked unhealthy and to give more light and space to those that remained. The trees were 6 to 10 inches in diameter at breast height, with about 8 to 10 metres of clear timber before the first heavy branches. I cut them to rafter length where they fell, so I could get at least two rafters and perhaps a stud length from each tree.

Extracting the stems from the wood to the waiting trailer proved to be an easy and enjoyable matter for the team of volunteers. Woven webbing straps and traditional timber carriers were used, with four or six people carrying each log. They made a winding trail through the trees, stepping over obstacles, up and down the ditches, causing no harm to the woodland. Probably less damage than a horse and certainly far less damage than a tractor-powered harvesting trailer. As Rob commented:

> *I was amazed that so much elm can be found relatively locally, that has grown to a size that can be used for something as important as the barn's rafters and that elms even get big enough to produce so much usable timber. And, also, how easily we were able to lift a felled trunk between four of us using manual grabber devices and how well we worked together as a team, even though some of us had not met many times before. As a result, we were able to extract a large quantity of wood with a few hours' work. Carrying hefty trunks of elm out of the wood in the middle of nowhere and eating the delicious home-made hot sausage rolls brought by Esther and Bruce (made from their own pigs) was a highlight of my experience of Barn Club.*

The final few days of milling turned out to be even more memorable. Graeme returned with his mill to saw the last of the timbers on the milling list. After the first day of clear weather and a steady breeze from the west, the wind turned overnight to the east and the skies darkened. We were about to experience a snowstorm sweeping in from Russia. Pete Branford, a wood-skills teacher and touring woodturner, summed it up:

> *Graeme was a hero. He was the guy who trailed his Wood-Mizer mobile sawmill, in the snow, all the way from deepest Essex, to snow-bound Hertfordshire and set up his machine in the teeth of a gale to saw up our trees. The 'Beast from the East' was fearsome, but our hero battled on with snow, sawdust and chippings being flung into his face. We even assembled a piece of fencing to shield him from the blast. It took a few days to appreciate the comedy of the situation!*

The temperatures didn't rise above freezing and the water-lubricated Wood-Mizer blade left ice over the surface of each milled timber. However, we were undeterred, and Esther Carley made sure that there was hot homemade soup and jacket potatoes to keep us warm. Bruce Carley kept our minds off the cold with hilarious stories of the exploits of his father-in-law's giant parrot who could apparently imitate a telephone ringing and whistle the theme tunes of old Western movies. Our motivation didn't wane, and, if anything, the experience of extreme milling drew the team to a greater strength of companionship. We worked on through the wind and snow, loading the mill and unloading the transformed timber. With the last rafter milled, Graeme bid his goodbyes and towed the sawmill away, leaving the sawdust and powder snow to finally settle across the yard.

With the timbers all milled and stacked, the scene was set for carpentry to commence. The following three chapters will relate in some detail the basics of how to mark out and cut a mortice and tenon for a traditional timber frame, all held together by a simple wooden peg. This will provide an insight into specific skills and knowledge that were once the bread and butter for village carpenters everywhere. These three simple things – the mortice, tenon and peg – were responsible for countless thousands of wooden structures throughout the centuries, of all shapes and sizes, including the Great Barn at Wallington and at Churchfield Farm. It is worth a page or two to describe the detail of their making and their significance.

EIGHT

# Two Buckets Full of Pegs

It's a wonderful and very satisfying thought that a structural frame can be made entirely with wood, human ingenuity and nothing else. An entire building, weighing several tons, including the roof, floors and walls, is supported on joints held together by pegs. No glues, bindings, straps, screws or bolts and only the occasional nail. Pegs are quick and easy to make with tools that can be carried around in a carpenter's bag. Although nails have advanced in design over the centuries, it is interesting that wooden pegs have stayed the same.

Pegs are made of good, straight-grained, heartwood oak. There are hundreds of pegs in a framed building, at least one in almost every timber joint, holding the joint together, but not in the way you might immediately think. Pegs come in different lengths, vary a little in width and are either approximately octagonal or square in section. The octagonal ones are tapered slightly from the head to the point, and the square ones are straight and true. The two types have very different purposes in the frame, as will be described later on.

The tapered peg is an integral part of the basic joint found everywhere in timber-framed buildings: the mortice and tenon. This joint has been used for centuries and the evidence is in the historic buildings spread throughout Europe. The idea of a tenon at the end of a post fitting into a depression or slot in a beam is an obvious way to prevent the two from sliding apart. Given that wood eventually rots, it is understandably difficult to find original timber frame joints from prehistory. But many examples exist from the more recent past, that is to say, from the last millennium. One of the oldest timber frames in England is the Barley Barn at Cressing Temple, Essex. It was built in the early thirteenth century and was itself adapted from an earlier, even larger building. Both its scale and complexity suggest that carpentry was already well advanced. We can only speculate at the scale of buildings and competence of the carpenters before that date.

As a school leaver I volunteered at the folk museum that had made such a lasting impression on me during my first visit. I was to help dismantle a seventeenth-century oak barn, in the weald of Sussex. It was being rescued from destruction resulting from a new road building programme. The whole building had been surveyed by Richard Harris for the Weald and Downland Living Museum. He had numbered and carefully labelled each piece of the sometimes severely weathered wood so that it could be removed, transported, repaired, stored and rebuilt in exactly the same form, but at a different site. Under Richard's beady eye, no piece of the frame was allowed to go missing or to be uncertain of its place amongst its sibling timbers.

By the time I arrived, the roof tiles and wall claddings had been removed, leaving just the timber frame. I was Richard's assistant for the week, and I can remember his arrival: his tall, lanky figure unfolding as he got out of his tiny car, a bright blue minivan parked next to the farm pond. It was summer and the frame was silhouetted against the clear azure sky.

The ancient structure of the barn was an impressive sight, though deflecting slightly along its ridge and one gable wall was not quite vertical. Over the centuries its original thatch would have fallen into disrepair and allowed rain to leak onto the timbers, resulting in areas of rot. It was looking a bit weathered and rickety, but it was all there and essentially the same as the day it had been raised 350 years before, lifted high into position in the air.

It was a glorious week of hot summer sun as we dismantled the frame, piece by piece. I was amazed at how easy and simple it was to take down this giant structure. We had a scaffold tower to work from and a pulley and rope to lower individual timbers to the ground. To dismantle the joints we used a ½-inch flat-ended round metal bar and a hammer. All that was needed was to give the end of each peg a sharp knock and the peg would work loose. Two or three strikes and it would come out. Then the tenon would slip out from the mortice and the timber would be free. Just like that. I held my first peg in my hand. It was a strange feeling. I had undone the moment in time, centuries before, when it had been driven into place. I looked at the thin length of dark brown wood in my palm, a peg that had held the joint in position for generations. It had held the barn upright through winter storms and autumn harvests, through the good and bad years of rural life. The barn had stayed upright in that spot for so many years. It was the expression of skills and intentions of the carpenters of the day and it was prevented from collapse by a couple of buckets full of wooden pegs.

The size of the peg was ¾ inch at its end and it had faceted sides. It was roughly octagonal in section as well. It was a tightly grained piece of wood, that is to say, very slowly grown. Sizes of modern pegs vary, but they are normally about ¾ inch at their heads. Pegs are made from green oak; oak whose cells are still wet with moisture and sap. Tightly grained oak cleaves more easily than fast-grown, so it is preferable. Because oak dries out so slowly from large pieces of wood, even logs that have been stored for a couple of years can be used. Dried oak can still be split, or cleft, but green oak is considerably easier: it just jumps apart when struck with an axe. Cleft oak has the advantage over sawn oak because it always splits along the line of the grain, and its fibres remain intact. The oak pegs also strengthen as they slowly dry in the weeks preceding the frame raising, dry oak being stronger than green.

Oak is used because its heartwood is our most durable and strongest timber. The tannic acid in the wood stains your hands purple-blue after a day of work, but the tannins also give very good resistance to rot and wood-boring insects. It means that oak is naturally durable and doesn't need any timber preservatives. Oak sapwood is not durable, so it is very important that none of it is used in making a peg.

Splitting fresh wood has an unexpected pleasure. You wouldn't know it from the dried timber furniture that we have in our homes, but freshly split green wood gives off a beautiful fragrance. Trees hold a hidden sweetness within. As you work with fresh timber, the smell drifts up. Wood smells as beautiful as apple blossom. You find you cannot resist inhaling deeply from a freshly cut piece of wood. Each species is slightly different, like the scents of different flowers. Each is a delight!

The log can be split with a regular axe into quarters. It takes a bit of practice, but it is worth the trouble to learn how to be accurate enough to land repeated blows in exactly the same position in a crack on the surface of a log, until it snaps apart. Metal wedges and a heavy wooden mallet, or 'beetle',

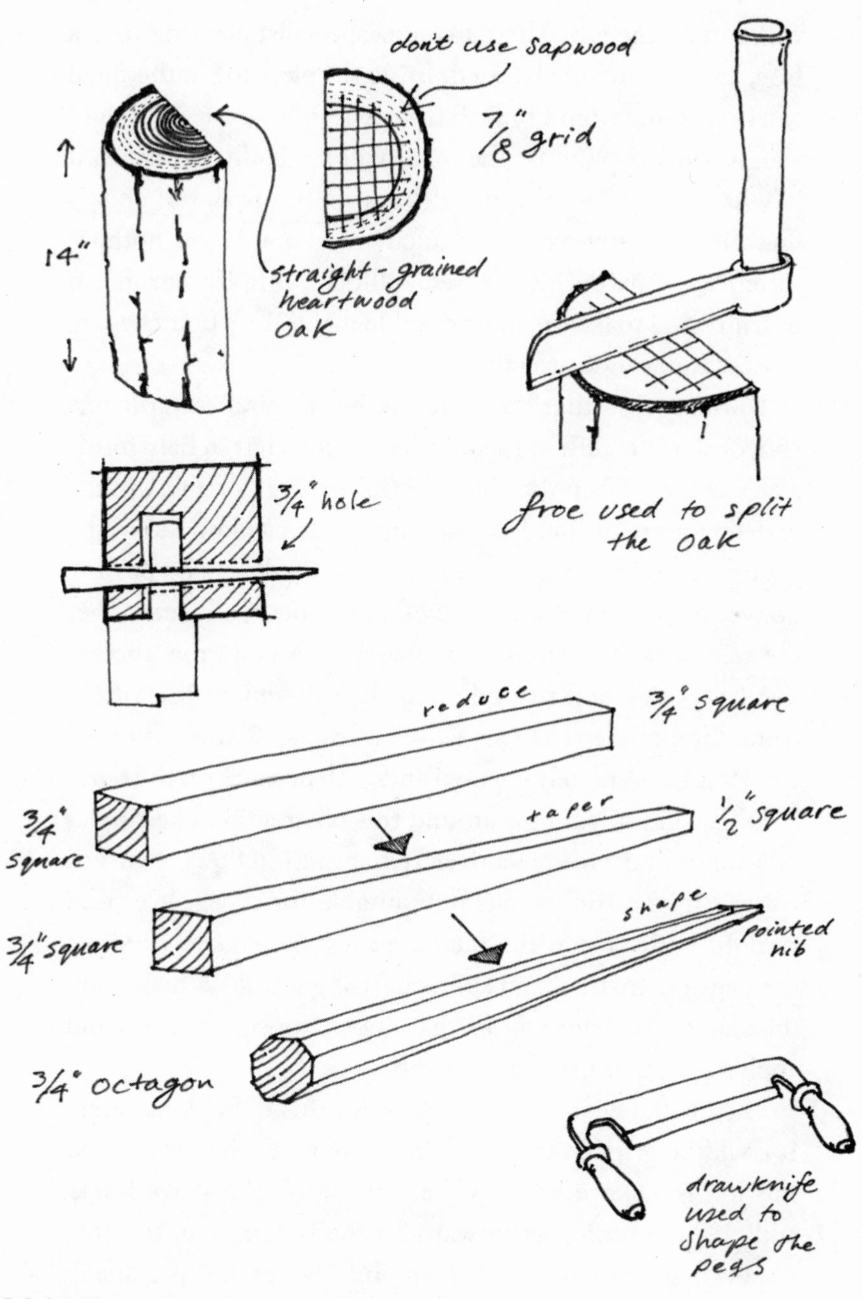


Making an oak peg.

will also do the job. Alternatively, a specialist cleaving tool, a froe, and a maul can be used. In its simplest form, the maul can be adapted from a branch from a deciduous tree. It should weigh about 2 pounds. The maul will get badly dented as it is whacked against the metal back of the froe, but this is unavoidable. However, the branch-wood maul costs nothing to replace. Any deciduous tree will do, but holly, hornbeam or fruitwood make the most durable mauls. Dry branches are best, as the wood is tougher.

You may find that the cleft oak has a spiral towards one end, or that the pith in the middle of a log cleft in half, forms a zigzag line. This pith is literally the sapling as it grew in the first few years of life. Like saplings in a meadow today, the young tree grows this way and that, pushed over by passing hooves or bent by the weight of a bramble. Only later does the tree begin to dominate its neighbours and grow successive layers of wood to straighten the subsequent line of the trunk. Sometimes the two halves of the split wood reveal a dead knot, completely encased and with no trace on the outer bark. The grain will flow around this obstruction like ripples of water round a reed in a river. It's a beautiful flowing curved pattern in the timber, but not suitable for a peg. We need straight grain. From the quartered log the square heads of the pegs are marked and split-out using a froe. When ready, the blank is transferred to a shave horse for squaring-off and applying a taper using a drawknife.

The blank needs to be held firm for this to be done safely. It could be gripped in a work-mate, but a far better tool for this job is a shave horse. When sitting on the shave horse, and pushing both legs forward on the bottom bar, the corresponding top bar bears down and clamps the peg blank with a force equivalent to a vice. The advantage over a vice is

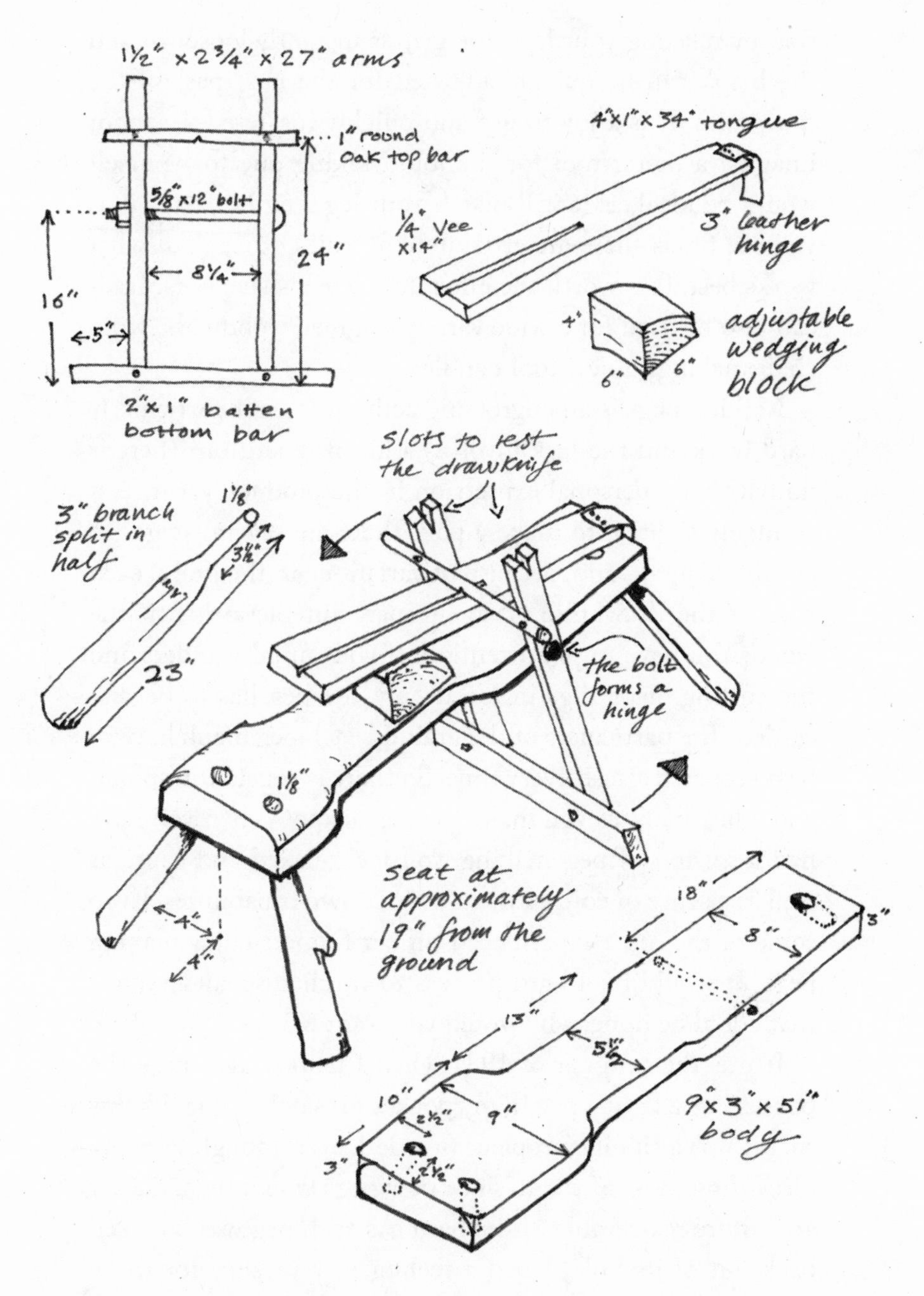


A shave horse for making pegs.

that by relaxing your legs, the grip is instantly loosened and the blank can be quickly adjusted for the next pass of the drawknife. It is a joy to use and quickly mastered. I cannot imagine a better tool for the job. Making one for yourself would be ideal, as it will match your leg length. Adapt it as you see fit, as there are absolutely no rules other than what works best. It is worth the effort to make a shave horse, as it can also be used for a wide variety of green woodwork, from chair making to new tool handles.

Making pegs is an engrossing activity. It's not particularly hard work and the task involves a lot of repetition. There is no particular personal expression in the product, yet it feels strangely creative to make a peg. There are several stages to get the hang of, and, because of variations in the grain, each pass of the drawknife is an intimate interaction with the wood. You have to pay attention. If your mind wanders and the cutting blade digs in too deeply, the peg has to be discarded. The particulars of the grain in each peg blank have to be discovered afresh every time. So there is constant response and adjustment as you make the peg shape. When you have had a session of peg-making, you feel relaxed and calm, as well as a sense of contentment at your own capabilities. Even some of the most experienced timber framers enjoy making pegs. It would be a hard process to mechanise, and even if that could be done, why would you want to?

It was the summer of 1981 when I helped dismantle the Sussex barn. I put a peg in my pocket for safekeeping. I knew that it was a significant piece of wood, even though we were discarding most of them, since new pegs would be made for any future reassembly of the barn. As each peg was knocked back out of its hole, I had a feeling of reverence for those long-gone carpenters. The ease with which we were able to dismantle the frame was a revelation to me. It was a contrast to the stability of the structure before we had started demolition. It showed me that this was not some crudely built thing that our ignorant forebears botched together, but rather a testimony to the care and skill with which they worked. I did not yet know that, although these rough-sawn timbers were irregular, they were jointed to an exact precision, and set to lengths to intersect to the precise, minute point of a carpenter's scratch awl. At that time I did not even know how to make a peg. But I had gained an admiration for this form of carpentry and a desire to find out how they did it. I had caught the timber framing bug.

So, why are the timber joints the way they are? What functions are the joints performing? The erected structure was stable and sufficiently rigid to withstand the test of time. Yet, the timbers came apart so easily the barn almost dismantled itself. There was obviously more going on than was immediately apparent. To understand this, imagine again a beam resting on a post. To stop it sliding about, a mortice slot is cut into the underside of the beam, leaving a width of timber either side: its shoulders. A tongue-shaped length of wood, a tenon, is cut on the end of the post to fit it. The weight of the beam is now bearing down through the shoulders on either side of the mortice slot onto the two shoulders on either side of the tenon. However, if someone shook the post, the beam could rock about on the shoulders. The obvious solution is to make the tenons inside the mortices very accurate and very tight. Strangely, this is the opposite of what used to be done. A far more ingenious solution was developed, and the demolished timbers of the old barn provided the clues.

By the end of the week's work on the old barn in Sussex, we had reduced the timber frame to a neat stack of tagged timber on the ground. The wood looked very weathered and insubstantial. It was hard to imagine the beautiful frame that it had been the week before. Looking at the stack end-on, each piece of wood revealed its individual tenon with a neat hole passing through near the shoulder. Some of the surfaces of the tenons were rough and not the sort of smooth finish I had seen in carpentry lessons at school. Some even looked as if they had been cleft with an axe or froe. This is where traditional timber-frame carpentry differs from joinery and cabinet making. In those traditions, the mortice and tenons are made tight and then wedged and glued into place. With timber framing, the tenons are deliberately made to be loose. The pieces of wood are so heavy and long that lifting them and pushing the joints together has to be done in one movement, rather than being held up in a precarious way whilst someone tries to bash the tightly-fitting timber into place. The same is true when cutting and testing the joints on the ground. When a big length of wood snags in its mortice, it binds horribly. It takes a lot of bashing and grunting to move it out again.

So, just as the timbers came apart with hardly any effort after 350 years, the frames needed to be assembled in the first place without the joints getting jammed. The secret is that the tenons are cut in a way that they are able to be in contact with only the shoulders and two of the five internal surfaces of the mortice. They universally bear on the back cheek of the mortice and the nose. The end of the tenon, as well as its face and the throat, are loose. Since these surfaces don't touch any others, they can be more roughly finished, as I had observed in the stack of demolished timbers in Sussex. Therefore, when the tenon is fed into its mortice, there is a lot of slop. They should easily slip into their corresponding

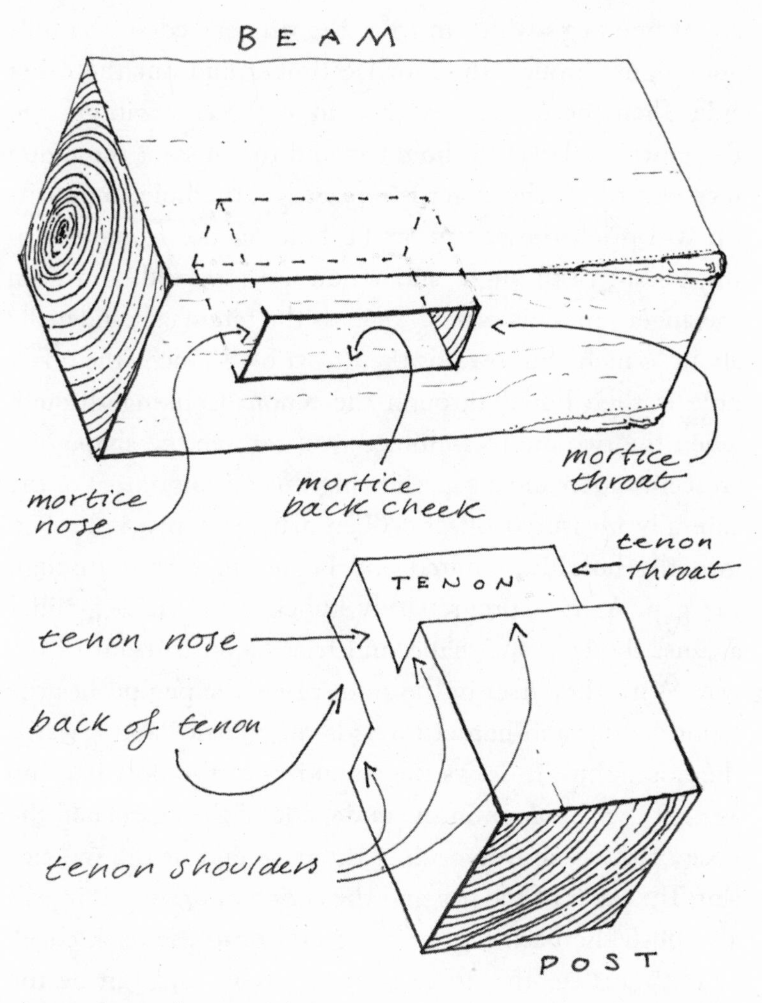


A mortice and tenon joint.

mortices, and the joint only tightens up when the pegs are finally driven home and everything locks tight – all thanks to the action of what is known as drawbore pegging.

The peg hole is drilled with a ¾-inch auger drill bit, but to start with only through the timber containing the mortice. The hole is usually in the centre of the mortice and about one peg's width in from the timber's edge. The hole goes right through the mortice timber and out the other side. Then the tenon is pushed in as far as it will go into the mortice, with the shoulders and the nose of the tenon nice and tight. The auger is replaced in the hole so that its tip will prick the centre of the hole on the front face of the tenon. Both auger and tenon are removed, and then the auger's mark is moved towards the tenon's shoulders by about ⅛ inch and towards the throat by 1/16 inch. The offset hole is then bored through the tenon. Looking through when the two are reassembled, you can see the shape of a crescent moon and a gap, showing this misalignment of the carefully measured offset. When a tapered peg is pushed into the hole, its pointed end begins to thread through the gap. As it is progressively knocked in, the peg slides against the crescent shape, and tries to push it out of the way. With the offset in the right place, the peg pushes the tenon harder and harder towards the mortice and towards the nose. Thus, it draws the shoulders of the post hard up against the shoulder on the underside of the beam, and the nose of the tenon up against the nose end of the mortice slot. The taper of the peg and the action of driving it in will also push the back face of the tenon onto the back cheek of the mortice. This force is considerable and can be the equivalent to one or two tons. It holds the post absolutely tight. It loses its waggle. The rule of thumb when pricking out the offset, is to imagine which direction you want the tenon to be pulled within the mortice and mark the offset in exactly the opposite direction. That will make the crescent shape where you want it when the peg is knocked in. The extremely tough cross-grained quality of elm means the mortice timber doesn't split under the massive pressure

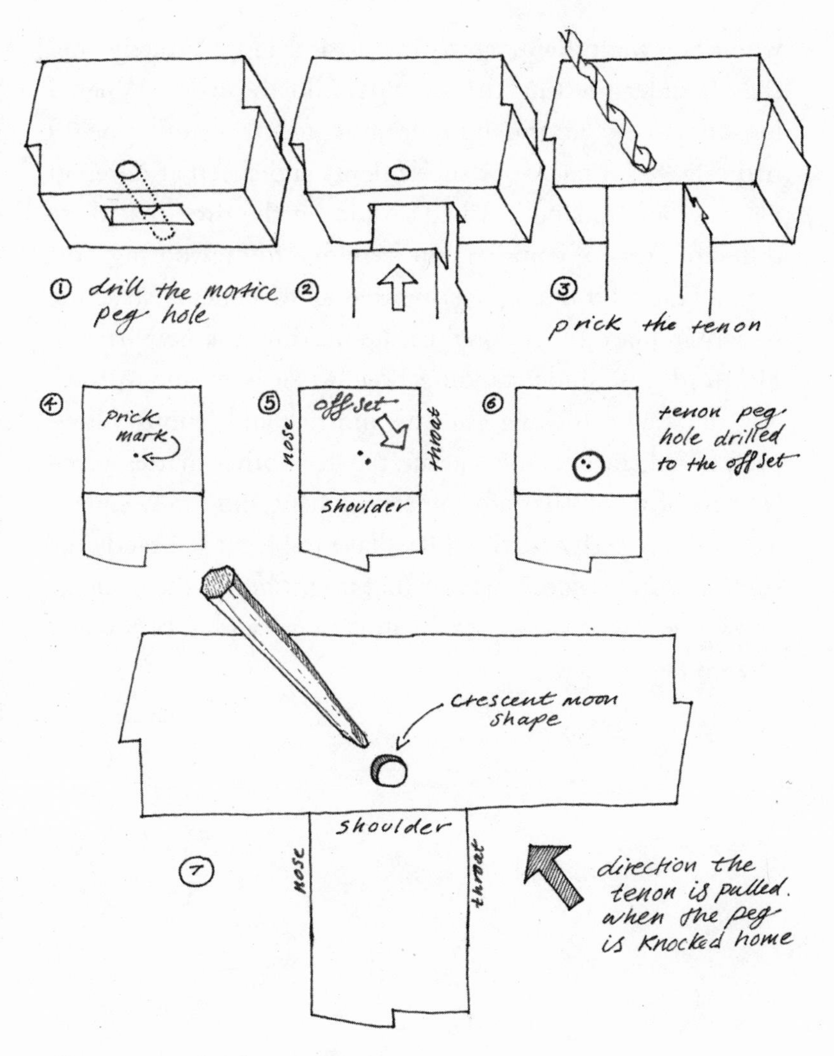


Setting out the drawbore.

of the offset peg alignment. Timber from many other tree species would split and the joints would fail if this drawbore pegging was attempted.

The hundreds of pegs in a timber frame tighten everything together as they are all driven home. Pegs hold the timbers where you want them, tightly in position and accurately onto the shoulders, letting the shoulders do the work. When I looked closely at the humble seventeenth-century peg I had salvaged, I could see subtle dents and sheen at different places along its length. I had no idea at the time that these different areas of compression were due to drawboring. It is a remarkable bit of engineering that allows a heavy structure to stay in place for so long. In the making of a new barn in Hertfordshire, understanding how a peg is meant to work was the key to making effective mortice and tenons. However, to cut the shoulders accurately is another matter, given lengths of wood that are sometimes bent and rarely square. To work properly, the shoulders have to be cut to exactly the right line and angle, whatever that might be. As the ancients knew, one answer to the problem is to use a plumb-bob and scratch awl.

The Great Barn, Wallington.

Two large healthy elms in Essex.

An elm-dominated landscape in Huntingdonshire.

Elm tree in a Cambridgeshire wood: a perfect forestry tree.

Native elm leaves from Hertfordshire.

The partridge breast feather pattern of elm wood.

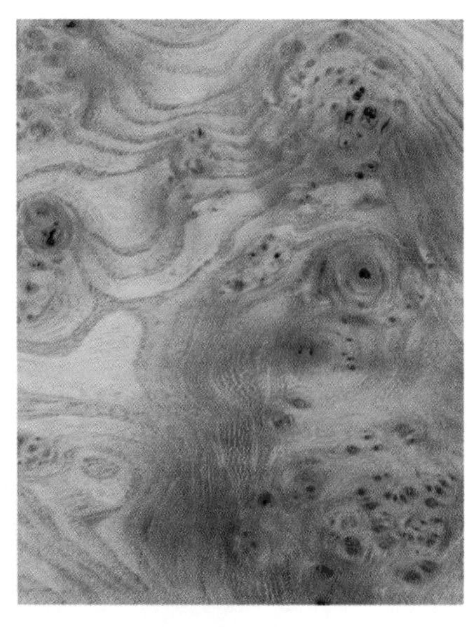

The swirling grain pattern of burr elm.

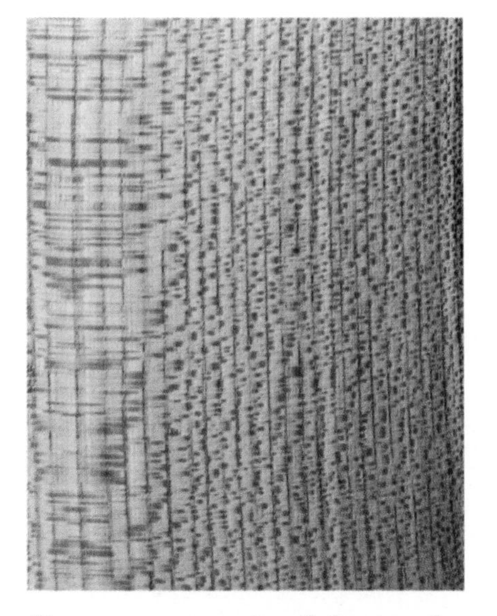

Quarter-sawn grain of elm wood showing the medullary rays.

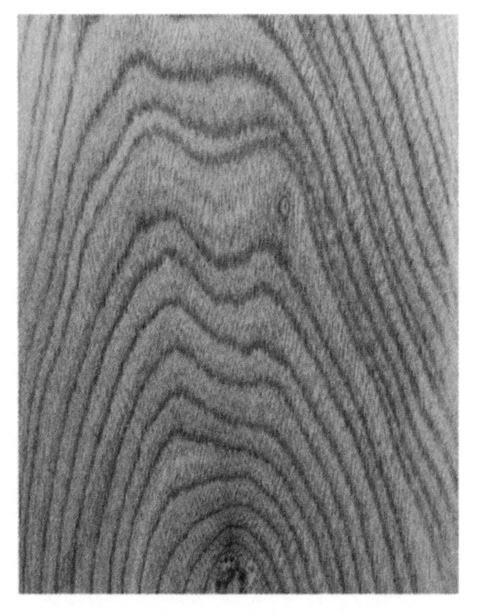

Crown-cut grain of elm wood.

The irregular heartwood-sapwood line in an elm log.

The elm trees felled and ready for extraction.

Elm logs milled to the sizes on the cutting list.

The freshly milled elm tie beam being wheeled out of a wood.

The mysteries of plumb-bob scribing being explained to Barn Club volunteers.

Sam looking along the line of sight to scribe a shoulder line with a plumb-bob.

Joel marking the mortice back-cheek using dividers.

Jess, Mihai, Pete and Bruce cutting mortice and tenons.

A busy framing yard with three cross-frame lay-ups underway.

Sean and Greg sawing the end off a jowl post with a double-handled saw.

Gable cross-frame-I completed.

Jo chiselling the shoulders of a jowl post to receive a top plate.

Rob ripsawing a tenon on a jowl post.

Esther shaping an oak peg using a drawknife and shave horse.

Cutting the rafter scotches together.

Jowl posts sawn, stacked and ready for the barn raising.

Marlin placing oak pegs in the peg holes of a top plate scarf joint.

The back wall frame in position ready to raise.

The back wall frame swings upright.

The post tenons guided into their mortices as the wall rises upwards.

Jowl posts upright and braced.

Both wall frames up and the sheerlegs in position.

Cross-frame-II tie beam and cross-wall in place.

Alex and Phoebe measuring the lengths for ceiling joists.

Cross-frame-II principal rafter being lowered onto a queen strut.

The final rafter swung into place.

The completed frame.

Barn Club.

Wavy roof rafters ready for fitting.

Fitting-out the inside.

The roof ready for tiling.

The store loft.

Pete and Zach working on the loft ladder balustrade and steps.

The main barn space and cross-frame-II.

Jo finishing the wooden handle on the entrance door.

NINE

# Plumb-Bob Scribing

A plumb-bob in its simplest form is a weight on a length of string. When it has finished swinging around, the taut line is vertical, perfectly vertical. Nothing could be more accurate, more reliable or simpler to imagine than this tool. Plumb-bobs vary in their appearance, and although it would be nice to have one that doesn't swing at all, they are usually made of something heavy enough not to be too easily blown in the breeze. The idea of using gravity to produce a perfectly straight vertical line wherever you need it, is another masterstroke in the evolution of timber framing. With the help of a pair of dividers, it allows us to make use of the most irregularly shaped timbers and still have tight joints.

There are many parts to this process, but once practised with real pieces of wood, there is a flow and logic that speaks for itself. The jowl post might look very complicated, but it is really just a series of mortice and tenons, each carefully scribed and cut, one by one, gradually developing into something that otherwise looks intimidating. A scribed mortice and tenon can also be achieved away from the workshop with simple tools in the field or wood. Just as learning how to use a plumb-bob properly is a liberating experience, so the application of the following instructions should release a deep satisfaction and springboard into all sorts of carpentry projects.

Let's return to the simple problem of fitting a post under a beam. The post has to be placed exactly where it is needed, perhaps as a doorpost for a particular width of door, and measurements on the framing drawings will show precisely where on the edge of the beam the edge of the post should finish: point 'A'. It seems a simple operation, but the problems arise when the sections of the timbers are not square. The shoulders on the tenon that will be cut out of the post head will have to meet the particular sloping surface of the underside of the beam. Added to this, the pieces of freshly sawn timber may well have slight curves along their length. Boxed-hearted stems hold their shape well when still green, but most timber that is through-sawn from small-diameter logs into slabs will distort across at least one plane. They may even be twisted slightly from one end to the other. You can appreciate why straight square conifer timber is prioritised for modern construction and home-grown broadleaved timber is normally consigned as firewood. But, back in the day, numerous ways were developed to get round these problems. What follows is one of these solutions.

Before reaching for a plumb-bob from the tool box, each piece of wood that is lifted from the timber stack has to be examined and assessed. The decision of where to place it in the frame will have already been made, and its sawn dimensions will reflect this. It will have been inspected whilst still in the form of a log and undergone a degree of visual grading starting right back in the woodland. Knots from branches and isolated rot pockets will have been noted whilst it was on the sawmill bed and then timber-sawn into a shape from the cutting list suitable for its quality or defects. However, for

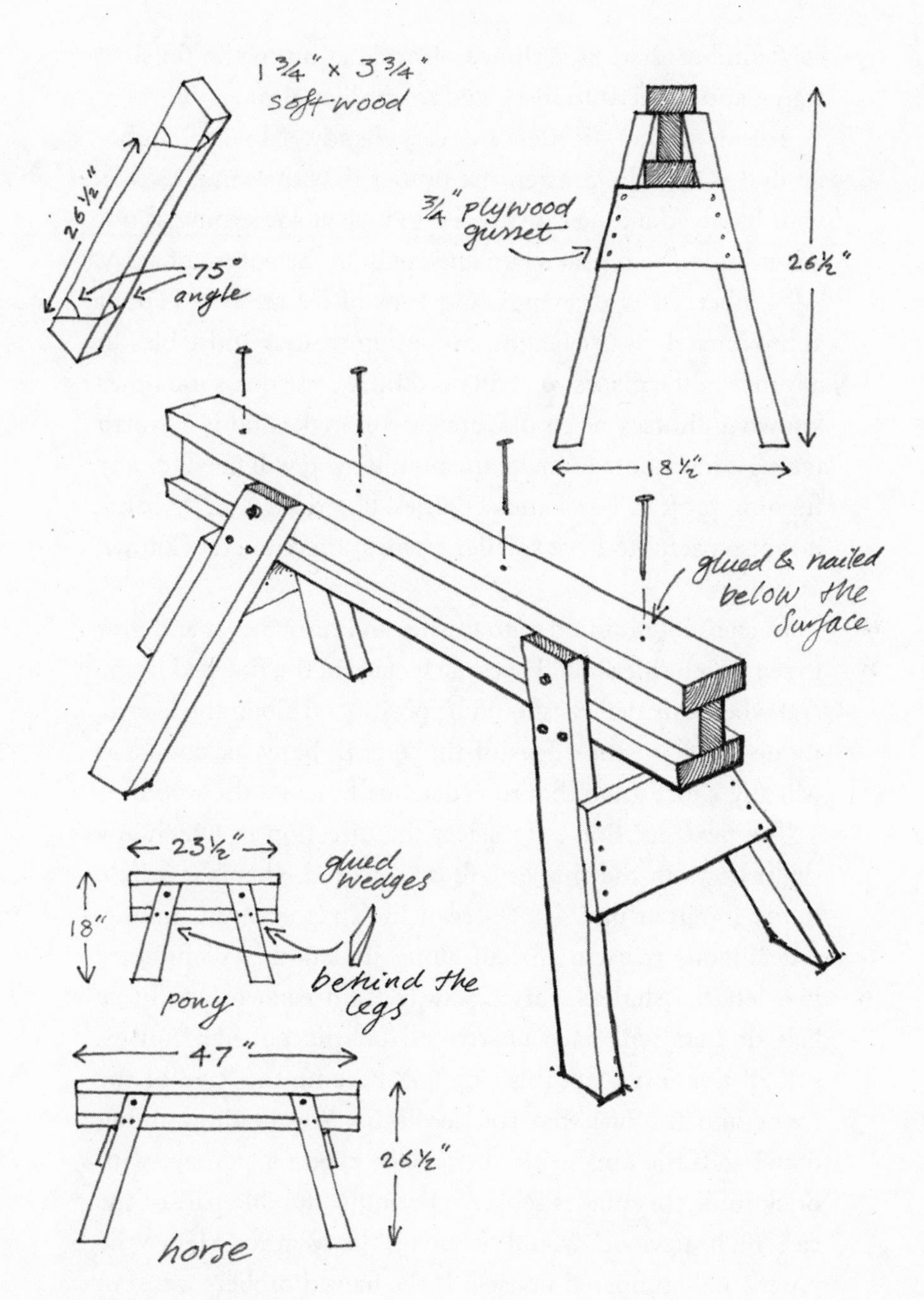


How to make a trestle.

each timber there is a choice of orientation in the finished frame and it needs to inspected from all angles.

This is where trestles are very handy. Moving timber around is best done when the timber is at the same level as your hands. Bending down, even to look at something, is not a good thing to repeat again and again in the course of a day. All timber, tools, drawings, and cups of tea are best kept at approximately waist height. So, set up trestles and tables in advance in the places you will need them. Trestles, sometimes known as 'horses' or 'stools', are easy to make and it is worth taking a few hours to make the number you will need for any framing project. For some activities, like ripsawing by hand, it is also useful to have smaller trestles, affectionately known as 'ponies'.

Each timber is lifted onto trestles and turned over and over to see which side should become its face in the finished frame and where any defects should be positioned along the length. Large knots on the edges of timber will be weak points, as will any 'short' grain that runs diagonally across the wood.

The next decision is to select the direction in which any slight bows in the timber will be oriented when the whole frame is raised upright. As seen in Chapter 7 on milling, a deciduous tree cut in half along its length will produce two lengths slightly curving away from each other. These half-timbers will have heartwood on the outside, convex, side of the curve. So, this side will become the 'face' of the frame and the one that you see facing you in the framing drawing. If the finished frame is to be exposed, perhaps with brick infill, this makes sense, as the most durable part of the tree, its heartwood, would be facing the weather. Hence the phrase 'half-timbered houses'. If the halved timbers were cut in half again, the tensions might result in a second curving of the wood as it came off the mill. The rule of thumb is to arrange the timber to have the heartwood outwards and upwards, so the bark will be on the inside or underneath. To a greater or lesser degree, each timber will bow up and spring out of the finished frame in the raised structure.

Finally, one edge of the face also locates the timber on the framing drawing. This is the 'face-edge'. So there will be a face and a face-edge side to each piece of wood. Once decided, marks are made to avoid future confusion. A simple fish shape is pencilled on the frame face with one of its tail lines carried on over the face-edge and down over the side. This identifies the orientation of the timber in the frame.

The task is to create a joint between the doorpost and the beam that achieves three objectives. First, the outer surfaces, the faces, of the two timbers must align to each other in the same plane. Second, the positions of the face-edges of the two timbers must also intersect at exactly one point, as located by a measurement on the framing drawings. Lastly, the shoulders must be snugly fitting together. This is achieved by accurate marking out and equally accurate sawing and chiselling.

## The Lay-Up: Aligned, Levelled and Plumbed

The first step is to bring the timbers to be jointed together, known as laying-up. By packing both timbers to become level and square to each other, they can be dismantled and reassembled again in exactly the same relative position. This is particularly important as more and more timbers are added.

A long spirit level placed exactly in the middle of the length of the timber will give a very good approximation of when it is level. Packers of small strips of different thicknesses of plywood are handy to put under the ends of the timbers

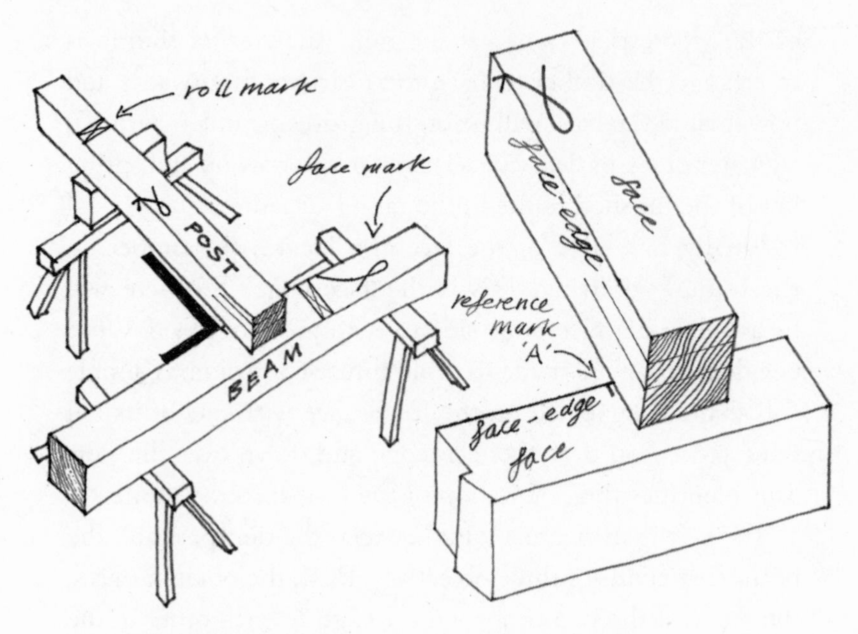


Post and beam laid-up.

to raise one end or the other to bring it to level. Also, wedges with a very low angle of incline can be useful if the timber is rolling from side to side on the trestle. A more accurate alternative, which applies if the timber has a pronounced spring in its shape, is to stretch a string line between the ends of the face-edges of the wood and to screw a temporary 2-foot length of batten so that it just touches the taut line. A spirit level can then be placed on the batten as needed.

That deals with the lengthwise levels. But the timber can also roll around its long axis. A small spirit level placed across the middle, at 90 degrees to its length, shows how level the face is in this direction. If the surface isn't flat, this can be trimmed with a small plane to make a localised flat surface. Wedges placed under the trestles can make the fine adjustments to bring the bubble to the middle of the lines of the level. A 'roll mark' is made at this point to show where to place the small level if the timber is moved and replaced. Initially, I draw two lightly pencilled parallel lines across the face, with diagonal lines between the ends forming an elongated 'X' shape. Later on, these lines are scratched permanently with an awl.

The timber to be morticed, the beam, is laid-up first on two supporting trestles. The face is upwards, and in this instance the beam is oriented so that the face-edge is towards the doorpost. The doorpost will be positioned at an allocated mark on the beam's face-edge, noted as 'A' on the illustrations. The face-edge of the post is positioned vertically above this mark using a plumb-bob. The doorpost tenon is to be 3½ inches long, so the post should have at least this length on top of the beam. The post's foot is supported by another trestle with blocks and thin packers on top to also bring it to horizontal. For this example, there is a right angle between the face-edges of the two timbers.

## Marking Tenon Lines

A feature of traditional timber framing in a barn is that any marks and scratches are left on the timber. They don't get sanded or scraped away. A barn just is what it is and has a rugged, honest beauty. So these various carpenters' marks will remain in place and help tell the story of the making of timber joints.

As described in the previous chapter, the nose and the back of the tenon will be in contact with the nose and back-cheek inside the mortice. The lines that mark the nose position and the back-cheek lines are the two critical lines in the marking out of both mortice and tenon. The approach normally taken

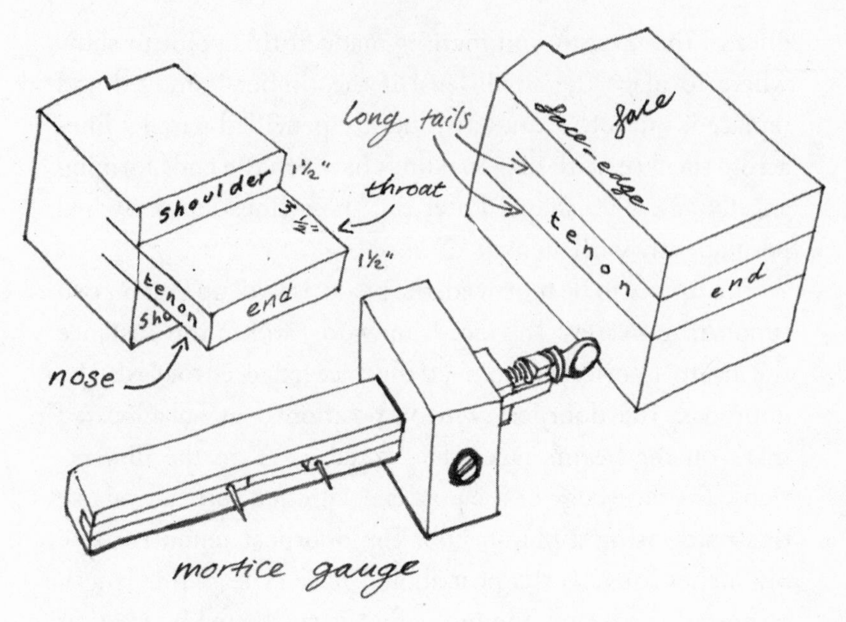


Marking the tenon lines.

is to mark up the lines for the tenon first and project these down onto the timber that will house the mortice.

Tenon sizes vary slightly according to the purpose of the timber they are joining. What follows is a set of dimensions that works for most joints. This tenon will be 3½ inches long, 1½ inches thick and be set 1½ inches down from the face of the post. These can be marked onto the post head using a mortice gauge. The gauge has two adjustable pins on its shaft and a head block. Set the pins to 1½ inches apart and 1½ inches back from the block. To mark out the tenon lines, the gauge shaft is held against the sides of the post and steadily moved forward, dragging the two points along behind. It works best to initially drag them gently so that they make soft marks and slowly rotate the gauge with each pass so that the points begin to dig in. Apply more pressure until clear scratch lines appear on the surface of the wood. Make the lines go well past the proposed length of the tenon. These long tails are important later on.

## The Drop

The 'drop' is a simple and brilliant concept used time and time again. It ensures the finished face-edge of the post will finish flush with the pencil mark on the beam. This is possible to achieve with extraordinary accuracy. Dividers provide this accuracy.

With the post in position and above the beam, the vertical distance between the face-edge of the post and the mark on the face-edge of the beam is the amount that the post will have to travel for the two points to meet in the finished frame. It is called the drop. In your mind's eye, the whole timber can be vertically lowered by the distance of the drop to end up where you want it to be. This distance can also be measured from the timbers themselves using dividers. The drop is set by placing the divider's points on the two face-edges that will come together when the joint is completed. It is worth noting that not only will the top face of the post be lowered by the drop to meet the top face of the beam, but also every other part of the post as well, including the tenon lines, will be subject to the same amount of drop. Through this, the exact position of the mortice can be marked by the dividers.

## Mortice Lines

So, take the dividers, set at the drop, and, working firstly on one side of the post, followed by the other, mark the drop from the lower of the two tenon lines above. The pinpricks can be circled with a pencil, or reinforced by using the scratch awl and then giving it a flick to one side. Prick and flick. The two prick points are then joined using a metal ruler to mark the line of the back cheek of the mortice. As with the tenon lines, the mortice back-cheek line is given long tails. As a reference line, the position of the back face of the post when it has dropped into position in the completed joint is also marked. It is pricked and flicked and marked with a dashed pencil line.

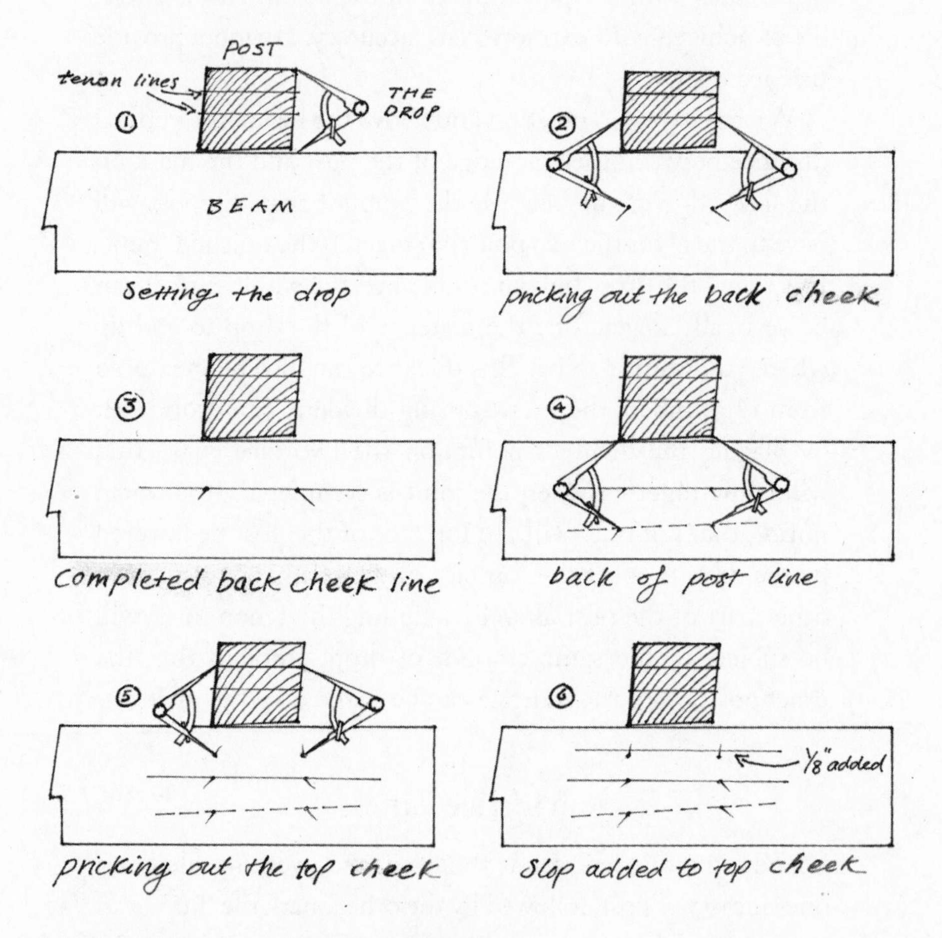


Marking the mortice lines using dividers.

When it comes to marking the front cheek for the mortice, there is another consideration. The tenon needs to be loose in the mortice, to make sure that it doesn't jam. It needs to have slop. So ⅛ inch is added to the 1½-inch mortice width. This measurement is set by using the smaller dividers set at 1⅝ inches, or by eye using the carpenters' square with one of its arms, which happens to be 1½ inches wide.

## The Nose

Both the tenon and the mortice now have their back and face lines marked, with nice long tails. The next stage is to mark the nose and throat positions, including any angle, or bevel, on the sides of the posts. This is where the plumb-bob makes an appearance. The plumb-bob string becomes a vertical reference line to measure from. A point on a timber at a certain distance sideways from the inside side of the string will be at the same distance from it after the timber has travelled the amount of the drop.

The plumb-bob is unwound so that the weight is just lower than the bottom of the beam. One hand holds the string on top of the post, whilst the other is free to steady the string and prick and flick with the scratch awl. You also need to position yourself so that your dominant eye can look either along the length of the post, or along the length of the beam or down the string itself. Each of us has a dominant eye. To find which it is, point to something in the middle distance and, whilst still pointing at it, close one eye and keep looking with the open one, then close that eye and look with the one that was closed. You will find that only one eye keeps your finger lined up with the object and this is your dominant eye – looking out of the other eye 'moves' the position of your finger.

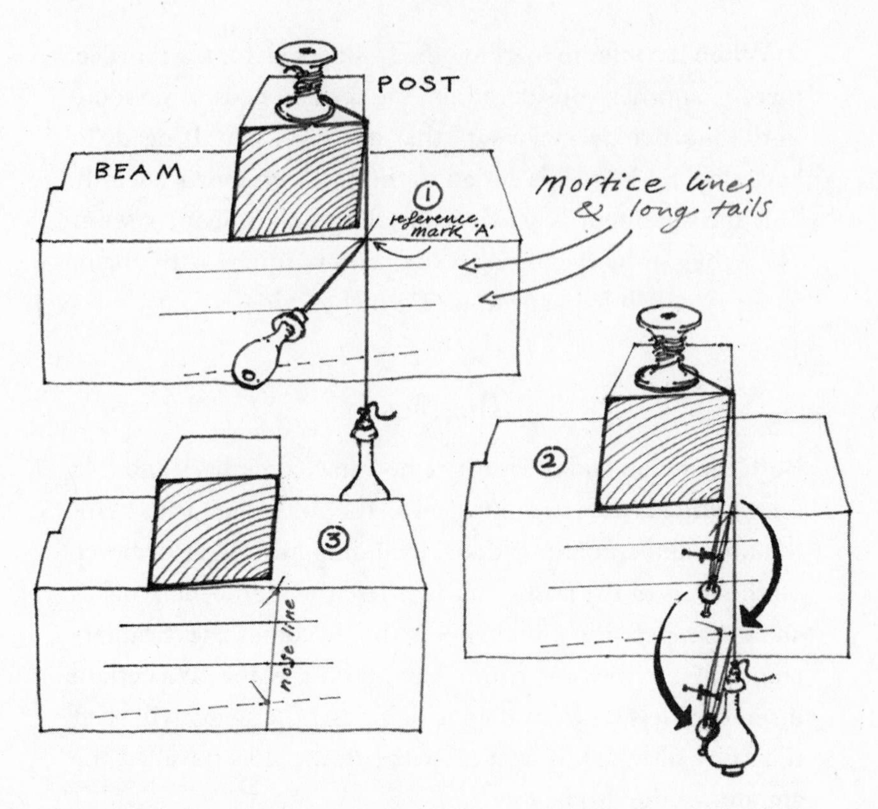


Marking the mortice nose line.

The plumb-bob string is held on the face-edge of the post and moved slowly along until it is just in contact with one of the beam's edges. Steady the swing of the bob. It is important that the string is not snagging on any edge beneath it. This is checked by moving your eye to look down the line of the string. If there is a snag, you will see a kink in the line of the string. It needs to be a completely straight line, just touching the edges. Look along the length of the post face-edge to observe the line-up of the string with point 'A' on the beam. The head of the post may need to be moved sideways by a small amount until the inside of the string is exactly on point

'A'. The awl is used to prick and flick a mark on the inside of the string at top of the side of the beam facing you. This marks the top of the nose line.

Looking along the line of the face-edge side of the post again, observe the horizontal gap between the string and the edge of the underside of the post. This gap can be measured with the small dividers or, with practice, by eye. Using the dividers, transfer the distance at the position of the post reference line (this is where the underside of the post will arrive in the completed joint). The dividers are placed with one point on the inside of the vertical string and the other point on this dashed line. Prick and flick. This marks the bottom of the nose line. The steel ruler is then used to mark a line through these two dots, giving a long tail as well. This line marks where the sloping side, or bevel, of the post will position when the joint is complete.

In this example, the side of the post is sloping away from the plumb-bob string line from top to bottom. However, it might equally be sloping in the other way, or occasionally the side will be completely vertical. The same approach applies. Use the small dividers to transfer horizontal distances between the string's edge and the timber, but always by the amount of the drop.

## The Throat

To mark the throat of the mortice, the plumb-bob is set up on the other side of the post. The upper and lower pricks are made as before to transfer the shape of the throat side of the post to the beam below. However, to prevent the tenon jamming as it is fed in, there is a gap on the mortice throat: another slop. So, at the final stage of scratching the throat

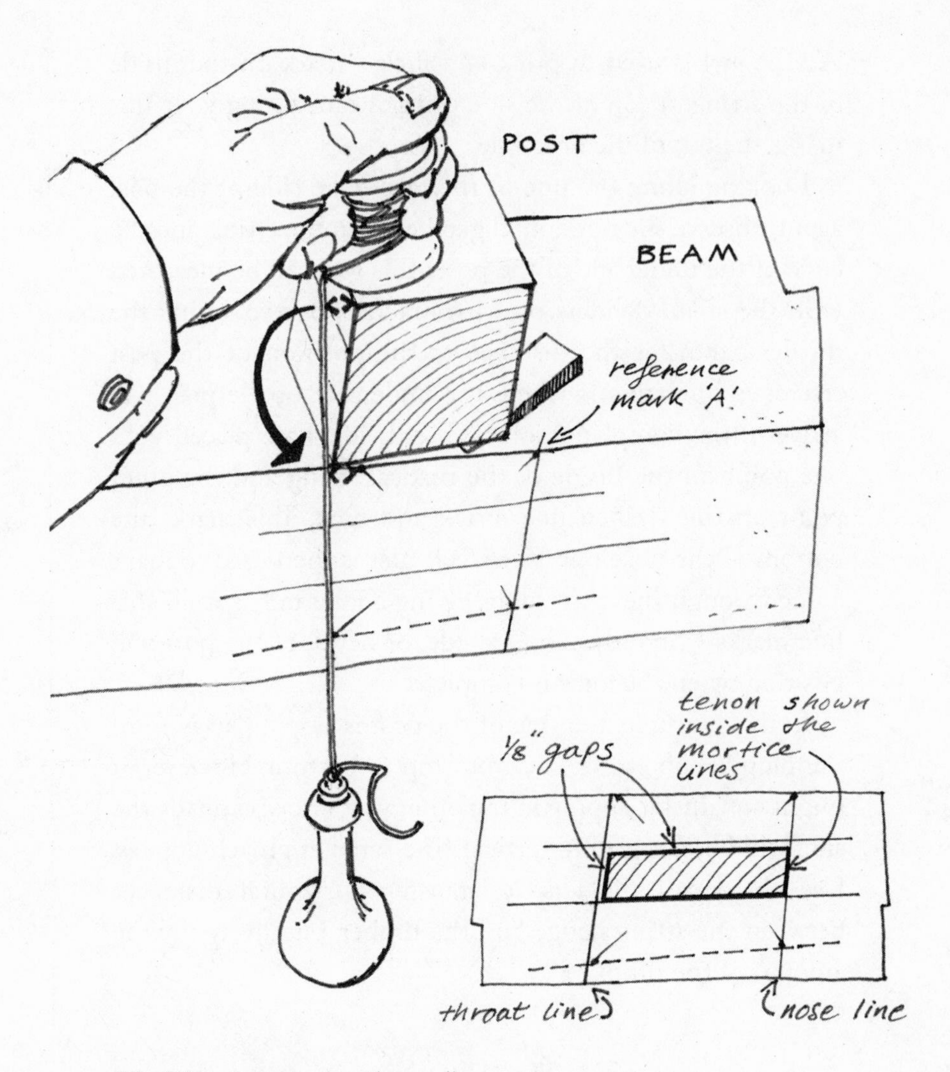


Marking the mortice throat line.

line, add a further ⅛ inch to the outside. This will give a little gap to the throat of the joint.

To finish marking the mortice, the scratch line tails can be gone over with a pencil, the corners marked with triangles and large circles sketched within the mortice box to show very clearly where the auger holes are to be made. See illustration 'All the cutting lines marked up' on page 181.

## The Shoulders

Next, the tenon shoulders. The two sides of the post are separately marked, working on one side of the post, then the other. Starting with the nose side, move position so that you can look towards the post along the line of the edge of the beam. Take care not to disturb the post where it rests on the beam. Hold the plumb-bob over the face-edge of the post until it again just touches one edge of the beam below. Steady the swing of the bob. Look along the edge of the beam towards the vertical line of the plumb-bob string. From this angle you can observe the slope, or bevel, of this side of the beam and can imagine this angle forming the shoulder line on the post now resting above it. As described below, you just need to transfer two points on this slope by the distance of the drop to the post above, and join the dots, for the bevel to be perfectly duplicated.

Firstly, the dividers are used to transfer the bottom point of the slope of the shoulder. This is where the dashed pencil line is again a useful reference showing the eventual position of the back of the post. The gap between the string and the beam at this position is the amount to transfer. Set the small dividers to the distance from the dashed line out to the string. Raise the dividers on the same side of the string line to the edge of the underside of the post. With one point touching the string and the other point resting just on the edge of the post, make a prick mark. Then use the awl to prick and flick.

Now to transfer the offset for the top point of the bevel of the shoulder. The dividers are set to the distance between the

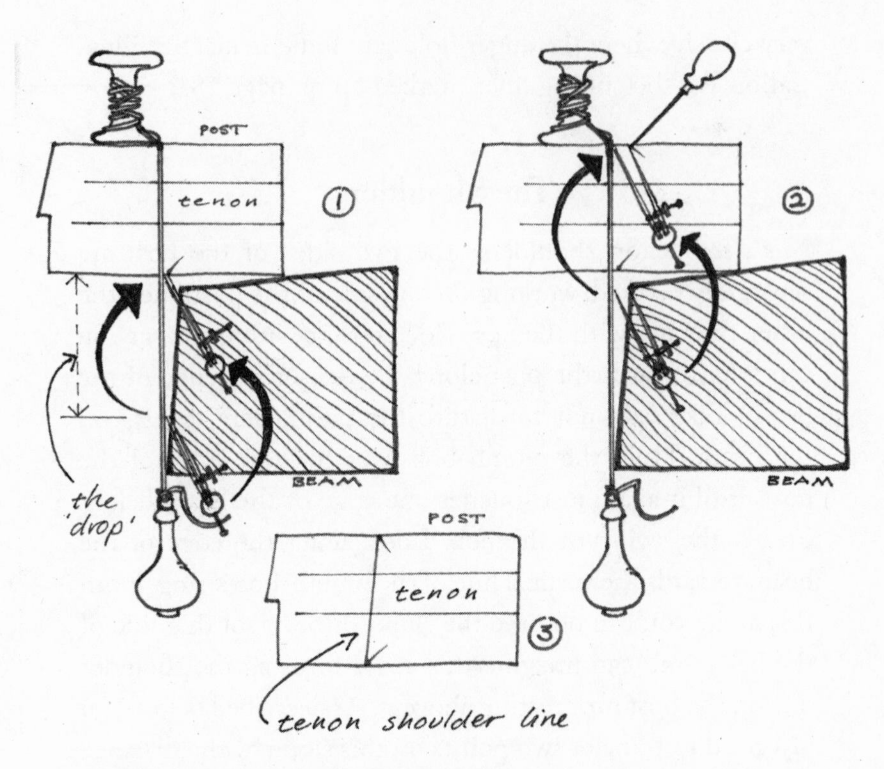


Marking the tenon shoulder line.

face-edge of the beam out to the string. This is transferred up to the top edge of the post, again with one point of the dividers just touching the side of the plumb-bob string. This marks the top of the shoulder line. Prick and flick. Lastly, use the ruler to join these two pricked points. This gives the line of the bevel of the shoulder at exactly the same angle as the side of the beam.

Move round to the other side of the post, and the shoulder on this other side is marked out in the same way. With the shoulder bevels now marked on both sides, the lines can be joined together across the top face. The letter 'S' is drawn, straddling the line, to mark the shoulder cut. Note which

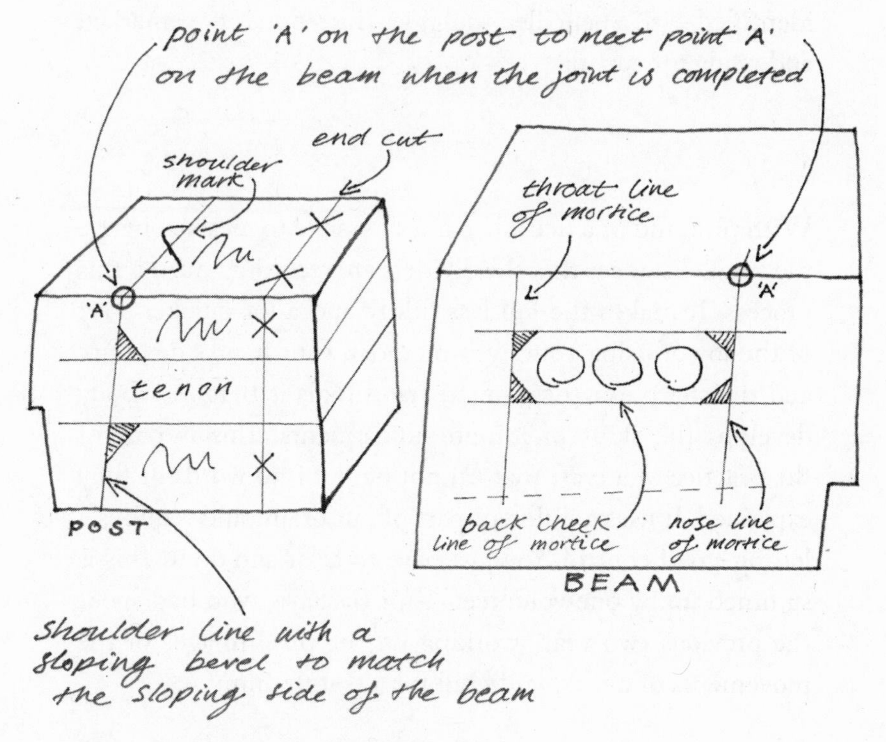


All the cutting lines marked up.

side of the tenon is to become the nose, and write 'tenon' in pencil along both sides of the tenon's tongue. The post can now be lifted away, flipped over and the last shoulder line connected across the back.

If the post is excessively longer than the tenon length, an end cut can be marked at 3½ inches along the tongue of the tenon. A line is marked that runs around the timber on all four sides at this length using a try square. It is marked with 'X's along the line to indicate an 'end cut'. For extra clarity, the corner between the tongue of the tenon and the shoulders should be clearly marked with a pencilled triangle. Finally, the areas of wood to be removed, the waste, is also identified with a pencilled squiggle. The tenon is now marked and ready for cutting.

---

With time and practice it becomes possible to measure by eye the small distances that the dividers are recording during this process. It makes the job less fiddly and a lot quicker. Part of the joy of using your eye's memory, your hand's dexterity and the sharp exactness of the hand tools is that your body develops the skills of minute adjustments. This is part of the practice of a craft that cannot be put into words or fully explained. It uses a different part of our brains and requires a letting-go of control. You just have to trust and do it. This is summed up by one volunteer, Tom Baskaya, who had spent the previous two years working on the pixel images of the movements of a mermaid's tail for a feature film:

> *For someone who works with computers for a living, this experience was something entirely different. When we were introducing ourselves at the beginning of the course I fell into the cliché of saying that I looked forward to doing something with my hands. It was politely pointed out that I would be using my brain too! Of course that was right. As I found out, I was about to use a part of my brain that felt like it had been neglected for a long time.*

One of the revelations of working this way, using dividers, is that tape measures are not constantly needed. Remembering numbers from a tape all day is mentally exhausting. Holding a figure in your head, then forgetting it, then memorising another one, then letting that one go, takes a surprising amount of concentration. At the end of a day your head is spinning. So, to have the dividers do the memorising is a big relief. Holding a plumb-bob perfectly still and transferring dimensions by means of dividers is strangely calming. It is an activity that requires concentration and an uncluttered mind. It is almost meditative.

If the marking of lines is a quiet and inward process, the cutting of joints is very physical and energetic. The contrast between the marking out and the sawing and chopping that follows afterwards is a lovely balance. That's part of the appeal. Lots of people comment on it because the emphasis on accurate marking of joints means that the job is as much a mental exercise as a physical one. However, sometimes you look at the lines to be cut with disbelief. They can appear to be at funny angles. It is worth double-checking everything before you lift the timbers apart, but, in the end, you have to trust your setting out and just go with the process. You won't know until the joint is cut and tested if your lines were right. And you won't know until the frame raising if all the timbers will all fit together anyway. There is a kind of tension in the air when the joints have been cut and the heavy pieces of wood are tested together to see if they fit. They usually do, and there follows a sense of satisfaction and relief. The whole process needs a kind of commitment and trust. So, with the mortice and tenon marked, the next stage is to reach for a handsaw, an auger, some sharp chisels and a good sound mallet to start cutting the joint.

TEN

# Cutting a Mortice and Tenon

The mortice and tenons on the Carley Barn were cut by hand. Power tools would have been quicker, perhaps, but would have made the project much more dangerous and unsuitable for volunteers. As it was, the team embarked on a steep learning curve to be able to saw and chisel with extreme accuracy. They all rose to the challenge. It always impresses me how quickly new skills can be absorbed and how much can be achieved in a day with lots of people doing a task together. As volunteer Rob Mills noted:

> *When I was first told to cut right to the scratch line, I didn't think it was possible. Inevitably, as a novice I was reluctant to do so. But with practice I found it entirely within my capability to do it and avoided subsequent fiddly trimming work.*

Hand tools, really good hand tools, give us a confidence in our own capabilities. So, when sawing, we shouldn't be fearful by sawing wide of the line. Aim to get it right first time and cut right to the line. After all, that's what all the painstakingly accurate marking-out is there for.

A sharp standard handsaw, one with eight to nine teeth per inch, is essential. For green, freshly felled timber, saws with fewer than eight teeth per inch also work very well. Blunt and crooked saws are a nightmare to use because they won't give the accuracy of cut. If you have a saw, turn it upside-down and look along the line of the teeth to see if there are any kinks in the metal or blunt tips, which show up as small points of reflected light. If so, buy a new saw. Disposable saws work fine, but they can't be sharpened, and kink and rust easily. For finesse there is nothing as good as an old-fashioned saw with an elegant wooden handle.

This degree of finesse in design is typical of antique, well-made tools and is often lacking in commonly available modern equivalents. Although it is possible to make a timber frame using standard general carpentry tools, the job is done more accurately and with much less effort using the old tools that had been developed over time for their specific purpose. They are a joy to use. If you find an old saw, as long as it is not kinked and the teeth do not run in a concave (upward-curving) line, they are worth sending to a saw doctor to get the teeth reset and sharpened. Afterwards, the teeth can be touched up with a triangular file without too much effort. Even large, two-person saws can be brought back to life. Once started in a groove, they are remarkably accurate and leave a nice flat surface. Some saws with seemingly ridiculously large teeth will work well on green wood, even if they are clumsy for dry timber. The basis of any sawing is to let the saw do all the work so that it glides. It's all a question of having a good technique.

To saw the tenon, lay the post on two trestles and make sure it is stable. Use clamps and thin wedges if necessary to stop any rocking. Both hands need to be kept free. This is really important. The timber needs to be held tight, but not by you. Holding the timber whilst sawing will result in less accuracy because you have to constantly lean on the timber using a lot of force. As you saw, you need to be able to freely shift your balance to be comfortable, or to steady the top of the saw blade with a free hand, or use both hands to hold the saw. You also need to be free to move your body to reposition your line of sight, which can be awkward if you are trying to hold down a piece of wood at the same time.

As a saw cuts across a piece of wood it leaves a cut, or kerf with a width of about 1.5 to 2mm. When cutting, don't cut in the middle of the line, which would make it disappear, but cut to one side, still keeping the kerf in contact with the edge of the line. Which side to cut? Always cut so that the kerf is on the waste side of the line, i.e., the side with the piece of wood that you don't want to keep. The marked line becomes one edge of the kerf as you saw. If it is a line scratched with a scratch awl then the line will be less than the width of a hair and the cut will be extremely accurate.

## The End Cut

The first task is to cut off any excess length of the tenon guided by the line previously marked by the 'X's. To start a cut, make a nick in the edge of the wood on the waste side of the line by lightly drawing the saw towards you. Move your head over the saw to align the length of the saw and the line you are cutting to. Look along the saw with your dominant eye. Use your thumb of your free hand to guide the side of the saw at the edge, with the side of your thumb just above the line of the saw's teeth. Make very short strokes with the saw to start with, and by leaning forward develop the kerf down the far side of the timber, rather than towards you where it is admittedly easier to see. As you saw, keep your grip on the saw very relaxed and just steer it where you want the kerf to gradually grow. Keeping the kerf exactly to the waste side of the line, extend it down to at least halfway down the vertical face of timber furthest away from you. Then, with the nose of the saw in that position, gradually saw the horizontal top part of the kerf back towards you until the saw is at about a 45-degree angle. You will have now set the kerf on the right track, on the waste side of the line. Keep sawing through, deepening the kerf at both its ends as you go. Finish with short strokes as the waste wood begins to fall away. With the excess tenon length sawn off, use the mortice gauge to continue the tenon lines around the end of the post. It is now ready for cutting the shoulders.

## The Shoulder Cuts

Begin the shoulder kerf again on the waste side of the line. Make a small nick as before and gradually extend the kerf down the far side of the post. The shoulder line should be touching the side of the saw's teeth as you draw the saw to and fro. Keep sawing down the far side to where the top of the two tenon lines is marked. Now, keeping the saw still, just touching the tenon tails, slowly bring the kerf on the top face back towards you, a little with each stroke.

It helps to think about creating a straight-sided slot in the wood, rather than cutting something off. Try and project your senses to feel the points of the teeth as they cut through the wood. Keep the teeth sliding freely, the less effort the better. Part of the purpose of the pushing stroke is to clear the sawdust. The sharp teeth and the weight of the saw will

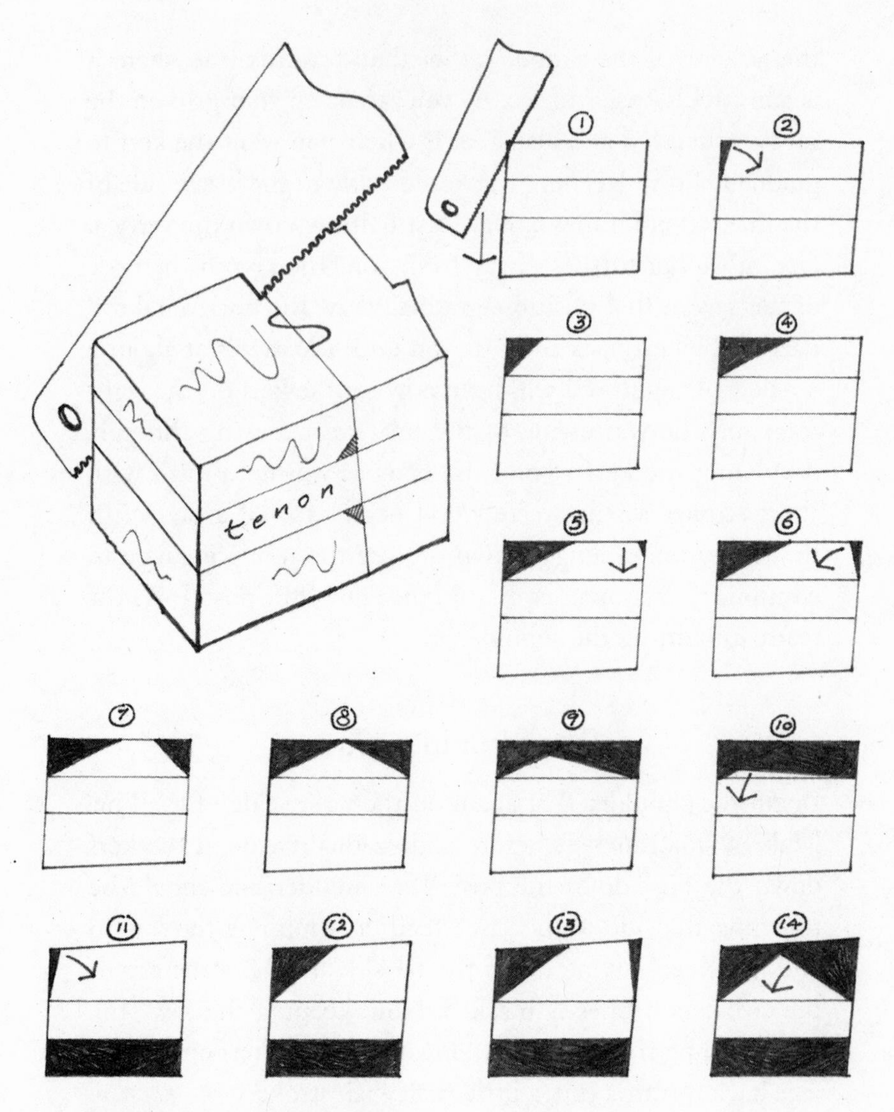


Sawing a tenon's shoulder.

do the cutting without much help from you. When you have cut halfway across the face, remove the saw and walk round to the other side. Repeat the process, again making an edge nick to begin. Lean forward to see down the far side and progress the kerf near vertically down to just meet the top tenon line. Extend this second kerf gradually back along the waste side of the shoulder line to meet the first. Then, whilst still keeping the saw's nose down in the kerf, extend the top of the second kerf into the first. Bring the two kerfs into one. Gradually saw in a more horizontal direction until you are sawing across the full face of the post. As the teeth get closer to the tenon lines, be careful not to saw into the tenon itself. Finish by using the pull stroke only to clear out the sawdust. Invert the saw to use its straight back edge to check if the saw cut now reaches both of the two tenon lines. The cut should be flat and there shouldn't be any rocking.

## The Rip Cuts

Now the back and front face of the tenon need to be cut. This requires sawing in the same direction as the grain of the wood, known as ripping. The post is turned over and re-clamped so that the tenon's tongue is vertical and in the direction that you want to saw.

Normal cross-cut saws make extremely hard work of this job: it's like sawing through frozen bread with a bread knife. The best tool is a ripsaw. Its teeth have a different shape, being more like a series of little chisels. They are perfect for ripping along the direction of the grain. These days, a ripsaw is an uncommon sight, which is a shame because they make such easy work of it. Fortunately, they are still available in old tool stores and can be brought back to useful life.

The cutting process is the same as before, but starting at the waste side of the end of the tenon lines. This time, cut the first near-vertical kerf halfway down before beginning to saw the kerf back along the tenon line towards you. When

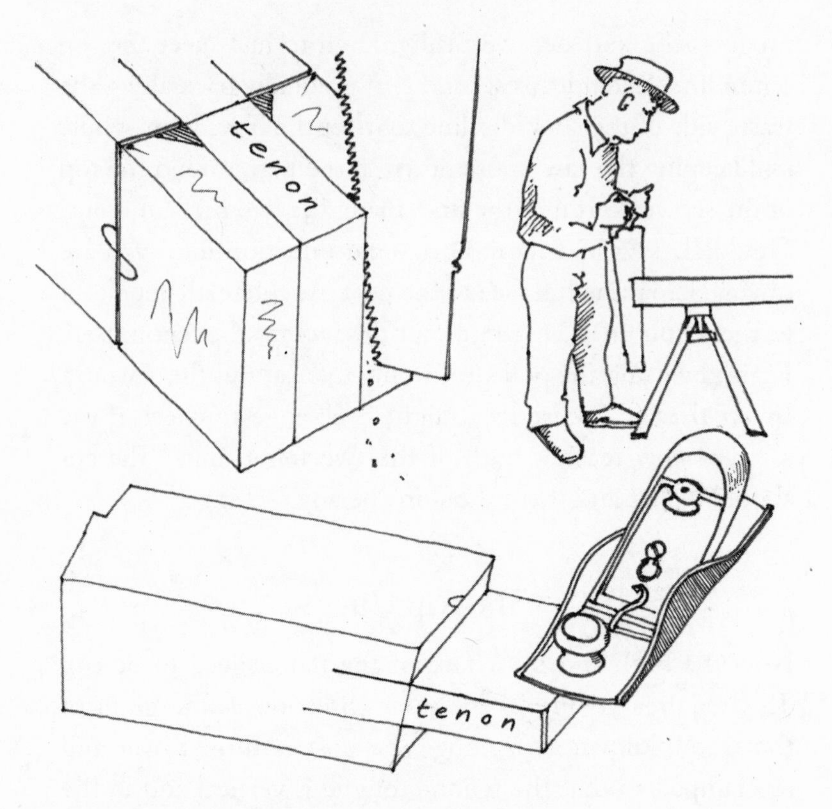


Starting and finishing a tenon's ripsaw cut.

halfway along this length, flip the post over and start all over again on the other side. Work the kerf down to meet the first one, then saw through both to the bottom corner before extending it back towards you. Because you are unable to see the kerf on the underside, it is worth turning the post over several times to gradually work towards the shoulders. The long tails help keep your eye in line.

Once underway, it isn't easy to change the direction of the kerf. So if it does begin to wander, lift the saw back to where the kerf was correct and, with gentle, short strokes, gradually form a new kerf in the correct line. Progress little by little as the teeth on just one side of the saw cut a new direction for the kerf. Gradually, the kerf will be back on track.

As the cut approaches the shoulders, a different handhold is used. Leaving the saw in the kerf, walk round the post to stand at its head. Holding the saw with a reversed handhold and leaning forward to look along its teeth, saw away from your body. Finally, the corner block of waste wood will loosen and can be knocked off with a mallet. With the second tenon line cut and waste knocked away, the surfaces of the tenon are revealed.

## Checking the Tenon

The straight edge of a ruler will show if the saw cuts are flat. If the surface is scooped out below the straight edge, then as long as the edges are accurate it can be left alone. If there is any rocking due to a slight humping in the surface, the hump needs to be removed with a plane or pared down with a chisel. To begin, use the corner of the chisel, rather than the full width of its cutting edge, to create a flat surface extended from the reference tails of the tenon lines. These corrections are gradually progressed into the side of the remaining hump, shaving strip by strip, until the surface is flat. Corrections can also be made this way if the edge of the sawn surface is slightly above the marked shoulder lines. Since the length of the tenon is 3½ inches, any surplus is marked and sawn off. To help the tenon slide into the mortice and not catch on its corners, a chamfer is planed or chiselled from the edges of the tenon's tongue. The shoulders are always left sharp and square. The aim of cutting a tenon with a saw is to cut straight, flat and true without the need to adjust it with a chisel. It's a real art, but once achieved, you feel really good inside.

## Augering the Mortice

Now to cut the mortice. To remove as much material as possible as easily as possible, an auger is used to drill out wide holes. These also create a clear space at the bottom of the mortice for the chisels to do their work without getting stuck. There are different devices to make the holes, and if you want to avoid using power tools, it is possible to use a hand-powered auger. This takes a lot of effort, but is how it used to be done. In the nineteenth century a hand-powered mechanical auger was developed in the US. These are still in circulation, and if you are determined not to use electricity, one of these will be very useful. Alternatively, use a high-torque, slow-speed drill. Although not as dangerous as a circular saw, they can cause trouble if the auger jambs. Your hands and wrists will get a nasty jolt if this happens. To reduce this risk, avoid using auger bits wider than 1¼ inches. Apart from this, it is a relatively safe tool to use with other people around. Considerably more dangerous is a chain morticer machine. These make a lot of noise and are not pleasant to use or to be near. The finished mortices that the chain morticer makes are rough and need to be cleaned up accurately with a chisel. The restrictions of the clamping system also reduce the natural shapes of timber that can be used. In summary, they are not as versatile as a simple auger and chisel. But they are quick.

To begin, the beam is turned back over so that the mortice is facing upwards. Again, the timber needs to be steady and firm on the trestles. An auger bit, say 1¼ inches wide, is used to drill a series of holes along the middle of the mortice. The end holes can touch the nose and throat lines, but not go over them. The holes are drilled ¼ inch deeper than the tenon, so 3¾ inches deep in this case.

## Mortice Chisels

To break out the rest of the waste wood, specialist chisels are needed. Mortice chisels are longer and stronger than standard carpentry chisels. They have heavier handles that are sometimes housed in conical sockets extending from the blade. The mallets can be wooden or metal with coiled leather pads on the ends to soften the impact on both the chisel's wooden handle and the wrists of the carpenter. Two or three pounds in weight is comfortable to use. An alternative is to make a wooden one from holly, hornbeam or fruit tree wood.

Sharpening is an extensive skill in itself, and a medium and fine sharpening stone are essential bits of kit in your tool bag. I use diamond stones and a sharpening jig to ensure an accurate bevel. It's easy to let the tool edges dull. When they do, little points or lines of reflected light appear on the tips when held in strong sunshine. The work also becomes arduous and the results less crisp. It doesn't take much to keep the edges sharp, or likewise to knock a chisel from its temporary rest on the beam. To avoid this, we used wheelbarrows lined with old towels to safely contain the chisels, planes and saws. The tools were thus kept near at hand, at the right height to reach and safe from being knocked to the floor.

The steel of a good tool will hold an edge for a long time. Whereas a tool that can't keep its edge isn't worth keeping. When sharpening, always keep the back of the chisel perfectly flat, only giving the bevelled-edge side a bevel. The most common angle is 30 degrees, which works very well with elm. A useful set of chisels would include 1½ inches, 1 inch and ⅜ inch wide. However, a single 1½-inch chisel will do most of the work.

Old tools often have very good steel blades. If they aren't damaged by rusted pockmarks, they can be reconditioned, resharpened and brought back into use. During the framing of the Carley Barn, there was a lovely moment one lunch break that illustrates this. Just as everyone was sitting down to eat, Bruce Carley produced an old hand plane. It was looking grimy, a bit rusty and had splashes of white gloss paint on it. Bruce explained that it was his Granddad's old plane. His Grandmother had saved up one Christmas and given it as a present. Bruce had inherited it, and, like many old tools, it ended up in the garden shed. Bruce had to admit it had seen finer days. Looking a bit neglected as it was, Joel Hendry, who was with us leading a carpentry course, recognised it as a Stanley No.3 smoothing plane. Rubbing the sides with wire wool soon displaced the grime and paint. Joel carefully re-bevelled the blade to a gentle curve using a wet-grinding wheel and honed the edge razor-sharp on a diamond stone. By the end of the day, the plane was restored to its former glory, much to Bruce and Esther's amusement. Granddad's plane became the go-to plane for the rest of the project. It proved ideal for quickly smoothing out humps on the faces of the timbers, leaving a clean, flat surface. Some tools are a pleasure to use, and this was one of them.

There is a strange and extremely useful chisel called a 'bruzz'. They are used for cutting the corners of a mortice. The really old ones have a V-shaped section, and these are far preferable to modern L-shaped corner chisels. The sharp corner of the 'V' bruzz gets right into the corners of the mortices without getting stuck. Unlike an 'L' shape, the ends of its wings don't dig into the wood. They pass through the free space of the auger hole, allowing the tool to be removed. Getting a chisel stuck in a mortice is something that you only want to experience once – performing the rescue operation can make a mess of the mortice. Using a 'V' bruzz was a revelation for me. They greatly reduce the effort needed to cut mortices by hand. If you get to use one, you will be equally astounded.

## Chiselling the Mortice

The following is a sequence that makes quick work of chiselling a mortice. With the auger holes drilled along the length of the mortice, use a ⅜-inch morticing chisel to knock out a channel along the middle. Then use the 1½-inch chisel to cut out slices vertically down on each side, leaving just ⅛ inch on the edges. Next, place the chisel exactly along the marks of the mortice line, and with the back of the chisel held vertical. Cut straight down to make a clean edge and a flat mortice cheek. Shavings can be scooped out with a chisel as you proceed. Sometimes the chisels make a wandering split along the line. This is to be avoided, but at least you have found out which way the grain is running. If that happens, move the chisel to the other end of the mortice box and start from that direction, so that the splits run into the waste wood rather than into the mortice shoulders.

The bruzz is now used to cut the vertical corners. Aim to get the exact vertical line of the corner. When working on the nose of the mortice, be careful not to go past the nose line. The bruzz can be knocked right to the bottom of the mortice, and as long as both its wings are free in the auger hole, it won't jam. Then use the 1½-inch chisel to continue the two sides of the mortice box right into the corners. Lastly, place the chisel across the grain at the nose and throat lines and chop downwards, being very careful not to

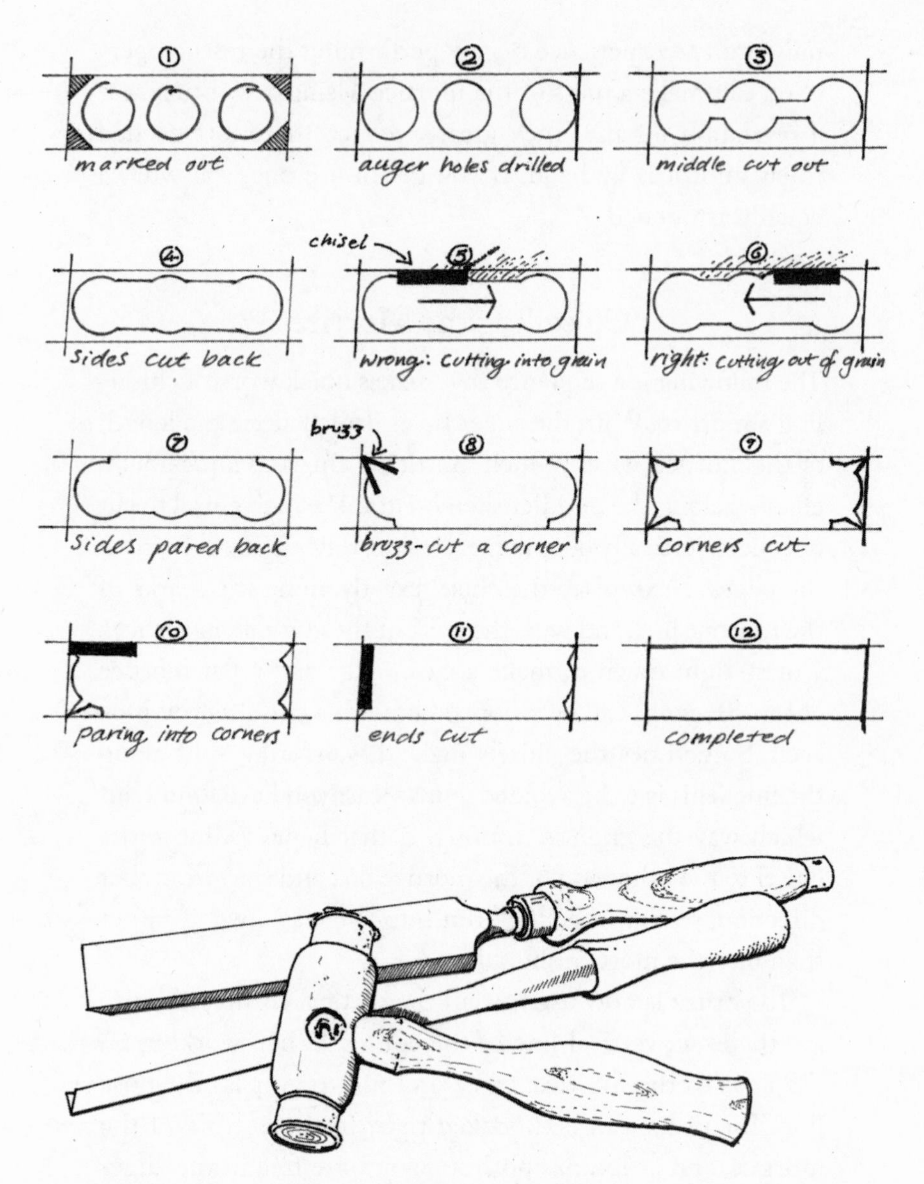


Chiselling a mortice.

overshoot the lines. Because the mortice was marked out slightly wider than the 1½-inch tenon to allow for some slop, the 1½-inch chisel shouldn't get stuck. The long tails in the marking are very useful to show where these final cuts should be.

## Checking the Mortice

There are likely to be bumps on the sides of the mortice, particularly close to the surface edges. Use a small adjustable square to test the sides. By holding a metal rule on the line of tails, you will see how much needs to be pared back. Paring can be done with any of the chisels, but the 2-inch-wide one is useful for sorting out the sides. A final refinement is to make the back face and nose of the mortice box undercut by about ⅛ inch. These are the two crucial sides that need to be cut precisely to the marking-out lines. This undercutting ensures that the tenon will not ride up as it is slid into place and it will bear exactly against the nose and back face lines of the mortice. The last of the shavings are scooped out, and the mortice is completed. The beam can now be rolled back over on the trestles to its original position.

Before testing out the joint with the tenon, the hole for the peg is drilled through the mortice. The peg hole is usually midway along the mortice and one peg's width back from the edge. For a ¾-inch peg, this means that the central point of a ¾-inch auger is set back 1⅛ inches from the beam face-edge. Using a drill or hand auger, the hole should be aligned parallel to side of the beam and square to its length. To ensure it is being drilled correctly, help from two other people will be useful. Standing in line with the beam and at 90 degrees away from it, they can sight-in the angle of the auger's shaft.

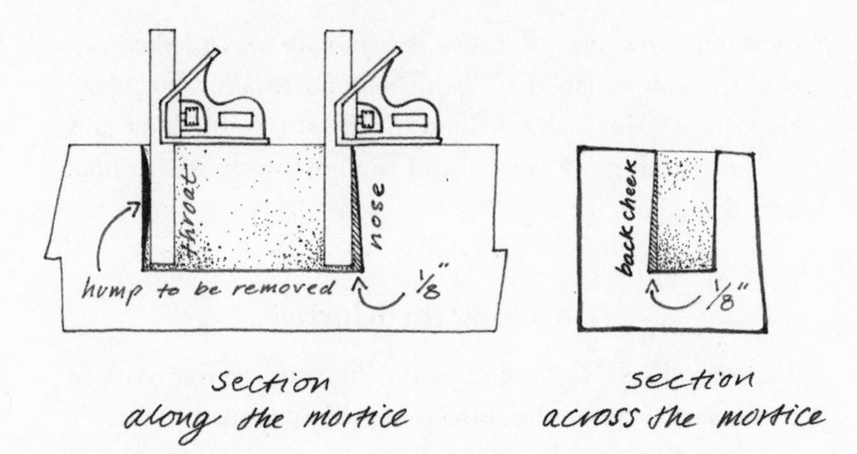


Checking a mortice.

Once drilled, clear out the shavings one last time. The joint is now ready to test.

Two trestles will be needed to temporally support the post in the right alignment before the tenon is safely supported in the mortice. If all goes well, the tenon will slide right up to its shoulders, closing up with a satisfying 'klumpf'. If so, happy days! The faces of the beam and post also need to meet without a step in the surfaces. Referring to the illustrations in the previous chapter, the two points marked 'A' on the face-edges need to meet exactly to achieve this. If there is a misalignment, additional paring down of the tenon's back face or nose will bring the two points into contact.

## Fettling the Shoulders

With the post now aligned how you want it, and at the required 90 degrees, the shoulders can be double-checked. With the tenon pushed firmly into place, there is sometimes a gap between one corner or more of its shoulders and the

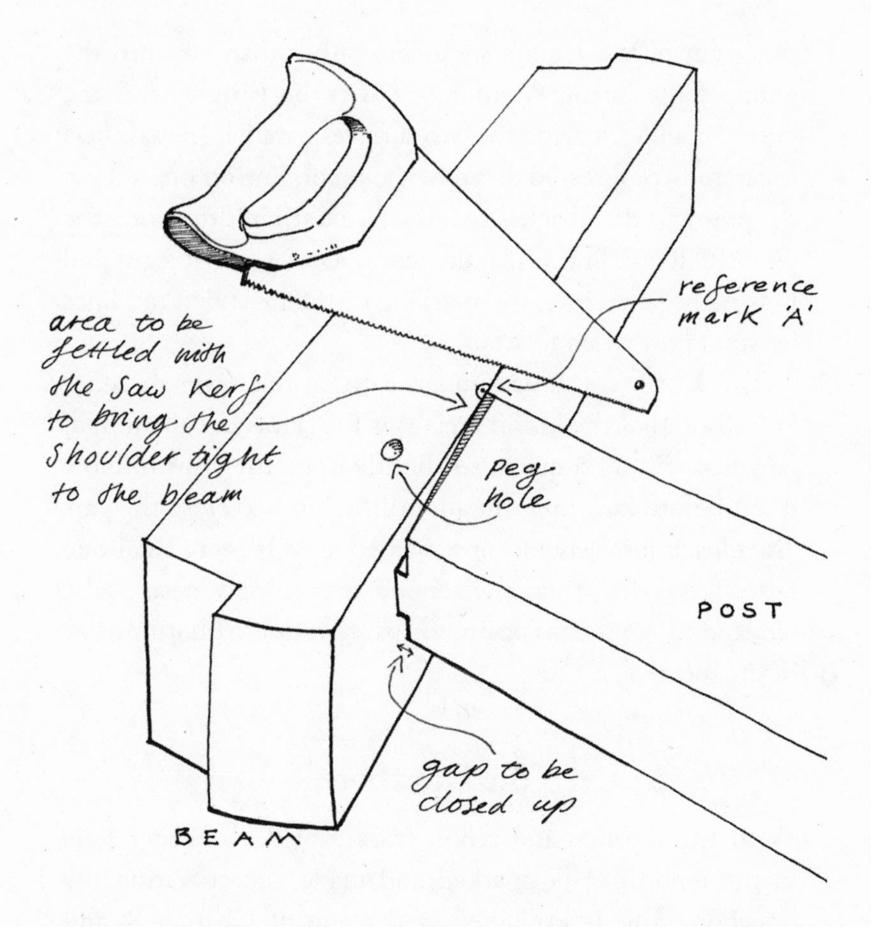


Final fettling.

beam. If three corners are tight and the length of the post is spot on, then it is best to leave the joint alone. Otherwise, as long as there is a little surplus length at the other end of the post, a useful trick can be employed to bring all surfaces of the shoulder into contact. It is called 'fettling'. A cross-cut saw is inserted in the gap with the side of the saw held flat against the side of the mortice timber. When pushed with short gentle strokes, the teeth will preferentially cut across the grain of the tenon's shoulders rather than cut into the grain of the beam. A smidgeon is to be trimmed off the post's shoulder, leaving the two surfaces parallel. The saw kerf is carefully progressed down to the tenon line on one side of the post and then back across the face until it is touching the opposite tenon line. Since the tenon itself is out of sight and mustn't be sawn into, the marking of the extended tail lines let you know when to stop.

This kerf is continued underneath on the lower shoulder as well, so that the kerf it creates is the same gap all around. Sawdust will be trapped, so slightly open the joint to blow it out before knocking the foot of the post to close the gap. The closed joint should now fit perfectly. If more than one fettle is needed, then the tenon's tongue may need to be reduced to 3½ inches again, to ensure it doesn't bottom-out in the mortice.

## The Drawbore

With the mortice and tenon reassembled, the auger hole in the tenon can be marked and made, incorporating the drawbore. This is explained at the end of Chapter 8, and the process is repeated for this joint. Temporary metal pegs called teepins, hook pins, drawbore pins or podgers, are used to test the drawbore. They are made round and taper to a point so that they can thread into the drawbore and put pressure on the 'crescent moon' of the tenon, which is visible when looking down the empty peg hole. Teepins are only tapped in gently so that they are easily removed. Gaps in the shoulders and nose will be closed up and the timbers brought into exact alignment. The joint is thus temporarily held tight whilst the rest of the frame is marked and cut. Eventually, an oak peg can be inserted and knocked home to make the joint permanent. Once knocked as far as it will go, the joint will be good for at least three hundred years!

---

With the completion of this one pegged mortice and tenon, explained here in great detail, the rest of the joints on the Carley Barn frame were mostly a matter of repetition of this process. There were slight variations and not all the joints were mortice and tenons, but the basic methods applied. In fact, the previous three chapters can be a springboard for the creation of all sorts of shapes and sizes of historic and contemporary handmade timber frames. If you can master the mortice and tenon, pegged with a drawbore, you are away.

For the Carley Barn, the ensuing laying-up, marking and cutting followed the tradition of Hertfordshire elm barns. Hundreds of mortices and tenons were scribed and cut by Barn Club members over the spring and summer. Each one invested with great care and attention, gradually building up a beautifully integrated structure.

ELEVEN

# Laying-Up the Frames

Doing the carpentry on-site, in the open air and right in front of the plinth where the barn is to be raised, is a very fitting thing to do. It feels right. Somehow it gives the frame an even stronger sense of place. There is a kind of beauty to the process as the timber frame gradually reveals itself in situ. It also feels strangely theatrical. The framing yard is the stage, the pieces of elm the actors. We, the carpenters, are the stagehands. Each frame lay-up is a different scene in the play. Each piece of wood has a character that we get to know as we handle and work with it.

Each 'frame', each flat plane that ultimately makes up the structure, is finished one at a time. For each frame, the timbers are laid out on trestles, with their ends overlapping ready to mark out and cut. Each of these is called a lay-up. In progressive lay-ups the joints are cut, the frame reassembled, more overlapping timbers added, more joints cut and so on until that particular flat frame is completed. The timbers are then put away and the lay-ups for the next frame begin.

Usually, frames are cut within a large industrial shed of some description and are churned out in the classic mode of a production line. They are then shipped, possibly hundreds of miles, to arrive and be hoicked into place with a crane. But where's the fun in that? It's like skipping to the last act of a play to save time in reaching the denouement. In contrast, Barn Club aims to maximise the richness of each stage of the process. The image I have in mind is of a medieval cathedral being built, surrounded in the cathedral grounds by the workshops and yards of all the different trades and teeming with people. In France there is just such a site where a castle is being built from local materials with hand tools in re-enactment of early medieval forms of construction. It is Guédelon Castle, near Auxerre, one hundred miles south of Paris, and well worth a visit. The really nice thing about the medieval approach to a framing yard is that you don't need any particular flat or even ground to work on. Just as long as it is not too soft underfoot, the framing yard can be wherever you want it. Each piece of wood will subsequently be levelled using blocks and packers on top of the trestles to accommodate sloping ground. So you can work out in the open pretty much anywhere.

There is a certain logic to the order of the laying-up. From the illustration of the system of frames making up a typical Hertfordshire barn in Chapter 4, it can be seen that each frame will include timber at its edges that also will appear in other planes as well. The sole plates form a horizontal frame on top of the plinths and they are laid-up again later on at the bases of the external wall frames. So they will take part in two framing lay-ups. Likewise, the jowl posts are fitted into both the cross-frames and the wall frames. The different frame lay-ups for the Carley Barn were as follows: the sole plate, each of the five cross-frames, the front and back walls, the top plate and tie beams, and the first-floor deck. Lastly, the roof structure was temporarily erected at ground level to mark and cut the purlins and wind braces.

## The Sole Plate Lay-Up

The sequence of framing inevitably begins with the horizontal sole plates that will support the rest of the frame. All the corners need to be jointed along with any longitudinal joints, known as 'scarfs'. Where I live, the elm frames traditionally have an oak sole plate. If there is rising damp oak will be more resistant to rotting than elm.

Individual lengths of sole plate will be brought out for subsequent lay-ups as the base timbers for all the wall frames. This meant they would have the longest time with their intricate joints cut and made vulnerable to drying and distorting. So, they were stacked in the cool shade of one of

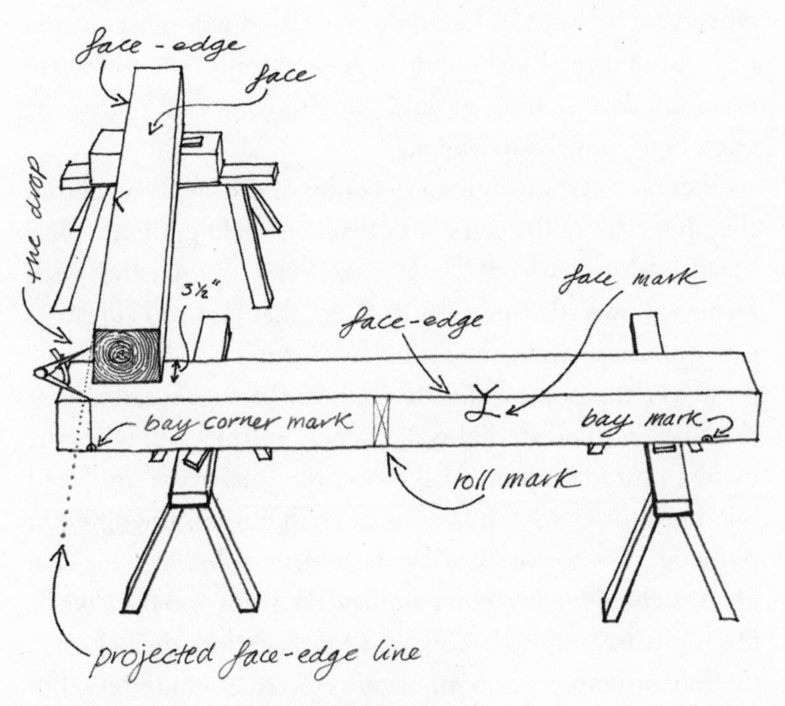


A laid-up section of the sole plate.

the two giant lime trees next to the framing yard. Sheets of ¾-inch plywood were also placed on all the stacks of jointed timber to keep the rain and sun away, which all helped to keep the timbers dimensionally stable.

## The Gable Cross-Frame Lay-Up

Having finished the sole plate, there is a choice in which frame to proceed with next. The top plates and tie beams could be marked and cut, or the front and back wall frames, or the cross-frames. For people new to timber framing, cross-frames are a good place to start. There was a strong desire amongst the volunteers to have an intensive week-long course at the start of the carpentry as an immersive introduction to all the techniques and tools. The advantage of starting with cross-frames is that several could be worked on at the same time, and that a week would be enough time to have something to show at the end. Credit and thanks should be paid to Joel Hendry, an experienced West Country carpenter, who came to lead the teaching. He carries the title of a master carpenter in all our minds. His unassuming manner, distinctive chuckle and enthusiasm for the work at hand was infectious. Herbert Russell, an historic building consultant and timber framer, came on one of the courses that were organised for that summer. He recalls arriving on-site:

> *Joel Hendry, the lead tutor on this course, had driven up from Devon to teach. Joel has a relaxed but authoritative charisma, and within moments of chatting, I could sense that he was more than happy to share his understanding of timber-frame carpentry. Other course participants then started to arrive and, as we were all chatting and laughing*

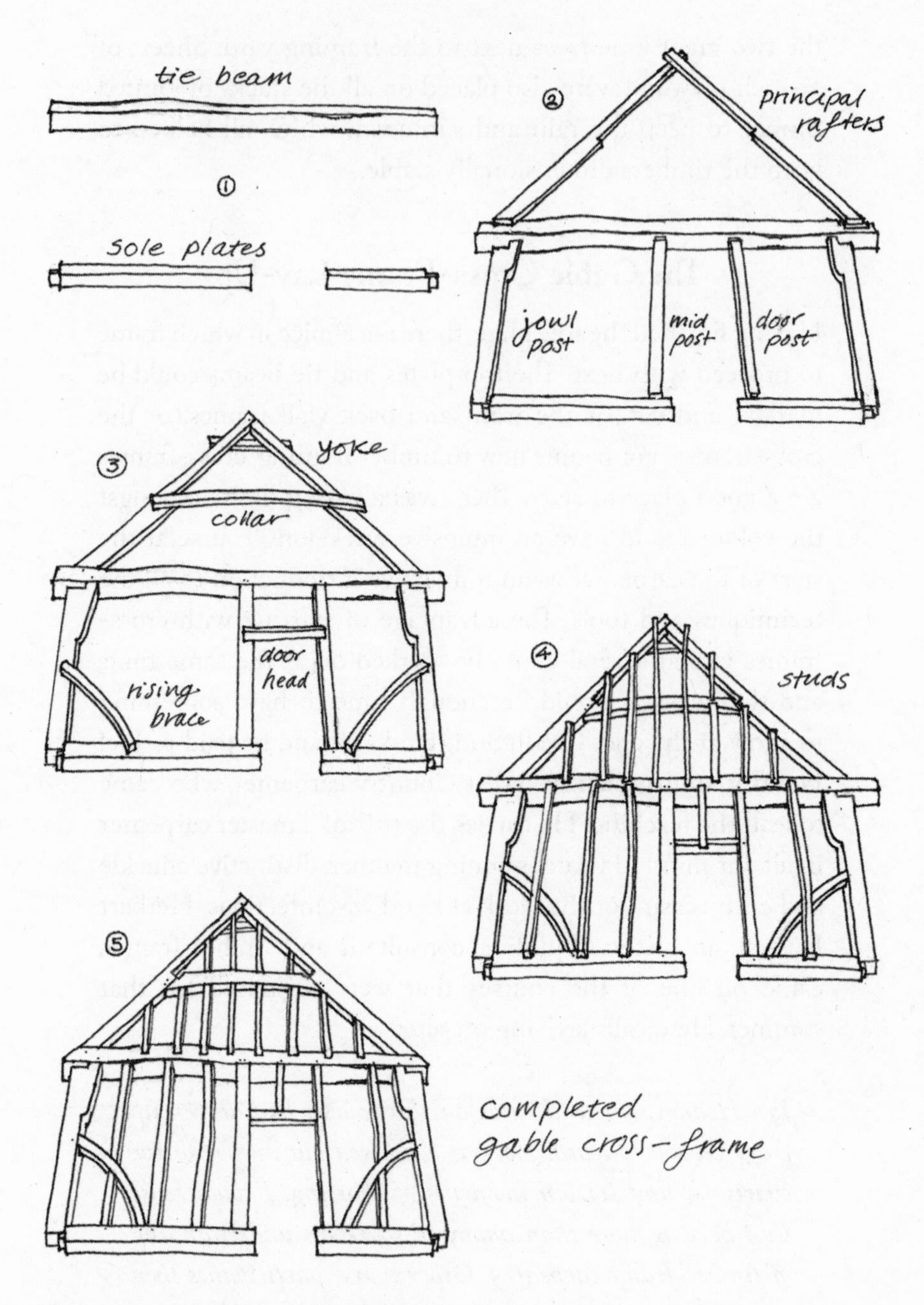


The sequence of lay-ups for cross-frame-I.

*whilst enjoying tea and biscuits, the team bonding and building process had naturally evolved. It is the little things that really count; the fun and inclusivity had already started before we had even touched the elm timbers or picked up a hand tool.*

For each cross-frame a tie beam is first chosen, inspected and oriented on the trestles. One nice feature of the open trusses is that the tie beams can undulate freely as they pass from one side of the barn to the other. The lines can follow the natural shape of a tree. Often, they are whole trees roughly squared with just one good fair face, perhaps an asymmetrical curve or hump along their lengths. On ancient frames, like the barn at Wallington, even the bark still remains in places after hundreds of years of use. These trees-as-tie-beams are fascinating, beautiful objects in themselves. They are usually the largest, longest timber in the frame. They are at the heart of it. Above all other timbers, they can retain the wildness of the woods, only slightly tamed by the marking and scribing of the framer. I always look for a stem that will make a beautiful tie beam in an open cross-frame. As Sam Rowland-Simms, a tall, amiable young carpenter who came to be part of the teaching team, observed:

*The thing that draws me in with traditional timber framing is how you never lose sight of the tree at the centre of it. The key question dealt with by the timber framer is, how do I accommodate this not-so-straight, often-twisted and 'imperfectly' formed tree into a structure that is upright, true and plumb? I struggle to think of a better combination of function and natural beauty than a completed timber frame.*

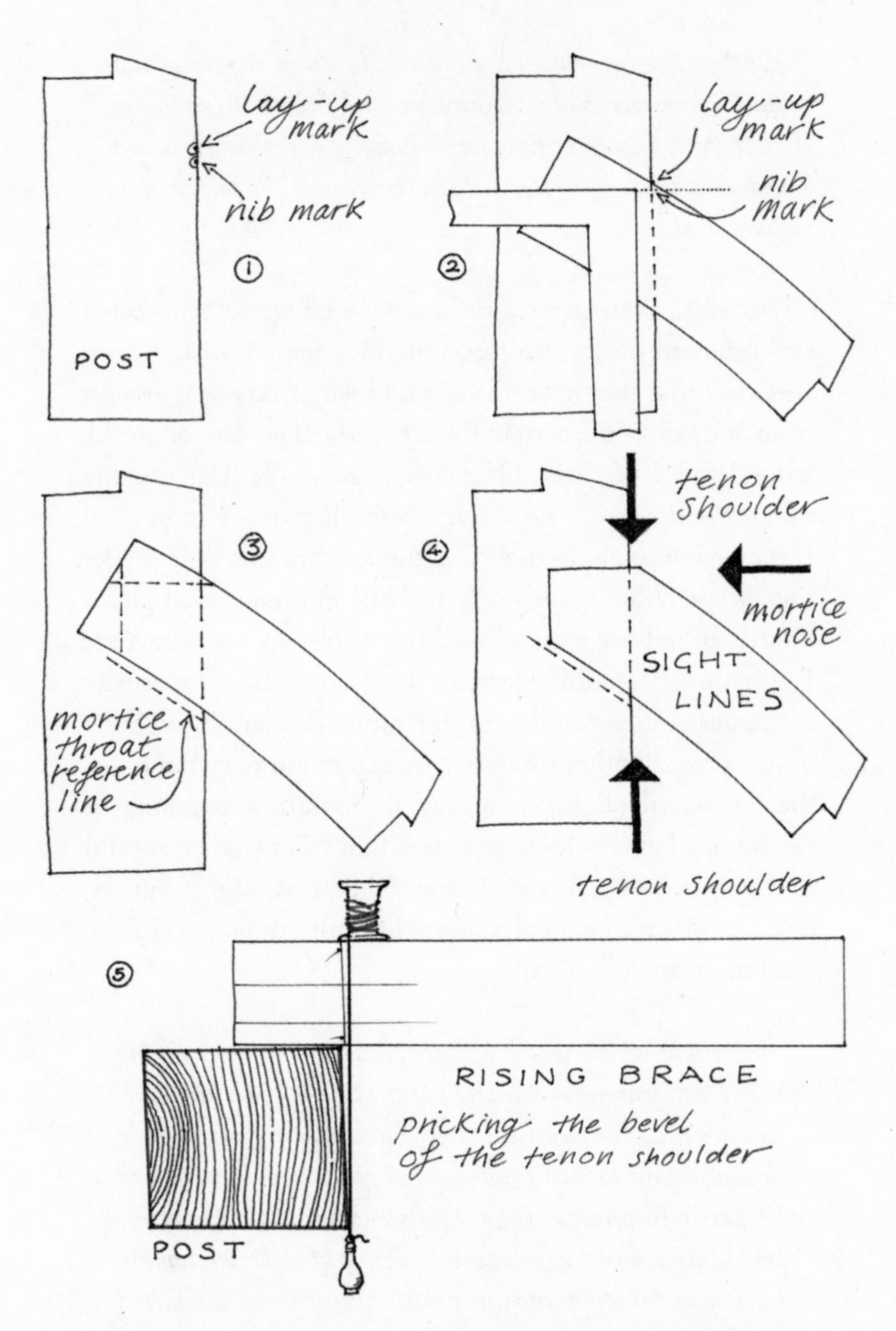


Pre-shaping the tenon and scribing the shoulders of a rising brace.

Once completed, the cross-frame with the collar and rising braces displays both the shape of a building and some gently curving timbers. It is a moment to stand back and admire the work. There is something about the balance of curves and straight lines in a timber frame that is very appealing. The clean arc of the curves gives a sense of movement to the overall shape. As Sam says, the combination of straight, squared timber, arching curves and natural undulations form a kind of harmony as you look across the structure.

## The Open Cross-Frame Lay-Up

In the open cross-frames there are no studs, and the expression of strength and grace as the timbers rise above you in the completed barn is wonderful. Key to this effect are the long arch braces, which rise from the jowl posts to support and triangulate the tie beam. Open cross-frames span across the open space of the barn, arcing high overhead and dividing the floor plan into bays. The number of bays might be just three, or in some cases can run to ten or more. By tradition each cross-frame faces towards the bay with the entrance doors, or 'mid-strey' as it is sometimes called. This enormous, roughly central doorway is a prominent feature that is common in most barns, and combined with the arcing pair of arch braces gives a humble barn a tremendous sense of grandeur. The form of structural arrangement in the roof truss above tie beam varies, with all sorts of collars, posts, struts, sling braces and purlins possible. They form beautiful shapes with hundreds of options, which are beyond the scope of this book. In Hertfordshire there are often just two queen struts that rise from the tie beam into the principal rafter, clasping the purlins as they do.

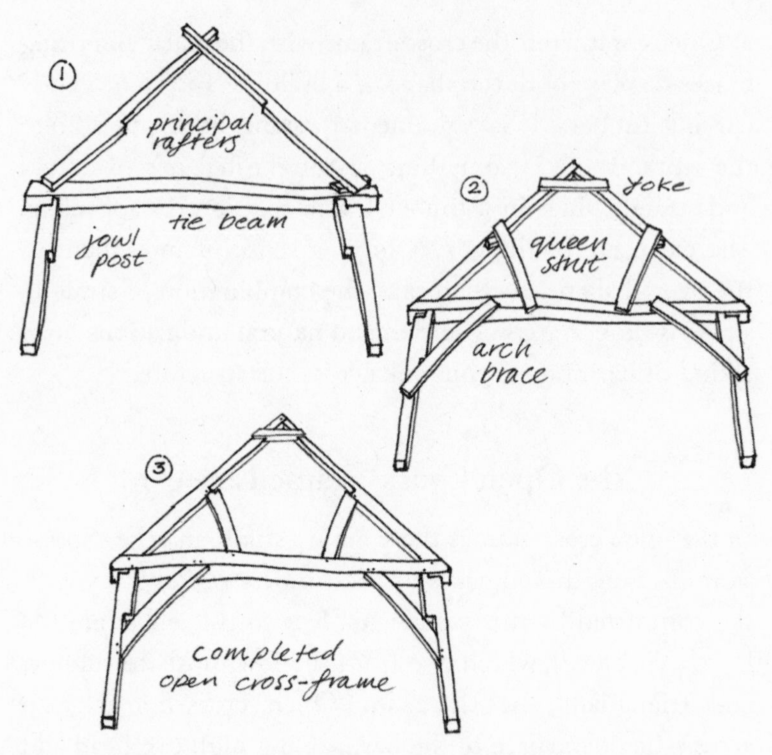


The sequence of lay-ups for cross-frame-III.

Arch braces are very important, both visually and structurally. So, the timber is carefully selected, ideally in symmetrically shaped pairs to match one side of the frame to the other. Arch braces are usually milled from gently curving bases of trees or large branches that are wide enough to mill a 2½″ × 10″ slab. These are milled an inch or two away from the heart of the tree to either side, to avoid heart shakes developing in the fitted brace. The curves are pencilled by eye or with the help of a flexible wooden strip that is held in place by small nails to induce a flowing bend that follows the grain. I cut these out with a large circular saw, though they could be cut with an axe, ripsaw or chainsaw. Arch braces

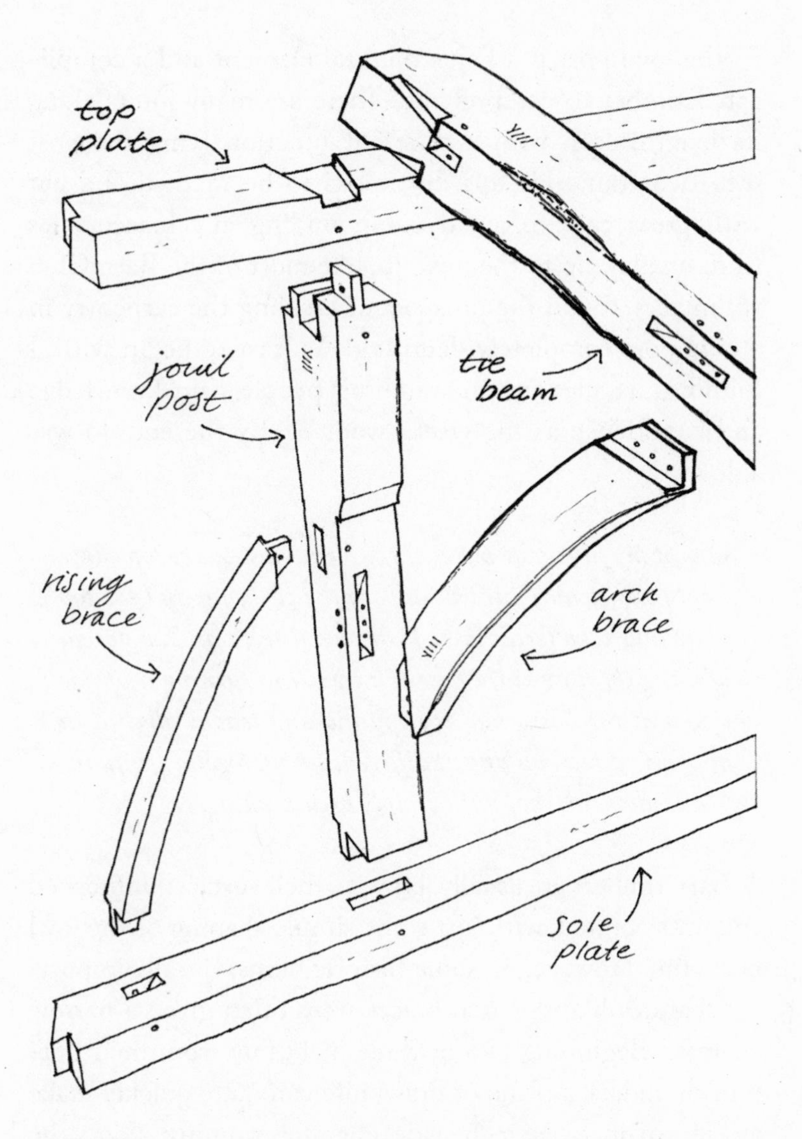


The jowl post and related timbers.

are often air-dried before use. This ensures that the uniquely long shoulders of the obliquely cut joints don't shrink and loosen as they dry.

The jowl post is a key structural element and a complicated timber to comprehend. There are many joints along its length, each with a different function tying the post into two frame lay-ups. So it has to be marked and cut with great care to avoid compounding any inaccuracies from one lay-up to the next. Jo, like most of the Barn Club volunteers, found the prospect of tackling the carpentry in a jowl post completely daunting. At least to begin with. I found great pleasure in watching people gain knowledge and confidence as the weeks went by. By the end, Jo was able to say:

> *One of my favourite pieces of the barn must be the top of the jowl posts, which joins with both the tie beam and the top plate and take forces in at least two directions. The notion of using the flare at the base of a tree, then turning it upside down as the basis, and then fashioning tenons aligned in opposing directions and at different levels, together with the 'secret' dovetail was quite simply breathtaking.*

Barn timbers are usually left with their surfaces unfinished and undecorated, with just some simple shaping of the jowl post chin. However, in some historic barns, visually important features like the arch braces were often given a narrow chamfer. Beginning two or three inches up from their bottom shoulders, a plane or drawknife can quite quickly make an approximate ⅜-inch-wide chamfer running along the long curves of the brace and out again at two or three inches from the top shoulders. Both downwards-facing edges of the brace would be treated this way. The chamfers are very muted and subtle in their effect, and you hardly notice them. But they do add something, and perhaps express the pleasure of the carpenters when they have fitted the braces and stand back to admire their work.

The last timber in this lay-up is the tiny top yoke that will support the ridge purlin when the roof is raised. Even though it might be just a scrap of wood, perhaps an offcut from ripping down the principal rafters, it is still carefully scribed so that it will fit snuggly into the frame.

## The Wall Frame Lay-Up

By late spring the carpentry for the Carley Barn was about halfway to completion. With the sole plates and cross-frames all cut and ready, the framing yard was surrounded by neat stacks of morticed and tenoned timber, each piece numbered with a carpenter's mark to record which cross-frame it was part of. The days were longer and the surrounding country was teeming with new life. It felt like a privilege to be working outdoors in the midst of it all. The pool of volunteers was also growing as word got round. Greg Cumbers came to join in. His keen attention and the sparkle in his eye spoke of his passion for working with wood and eagerness to learn. A former life as a touring musician had taken its toll, and he was searching for a more meaningful way of making a living and being creative:

> *I'll always remember my first few weeks volunteering on the project. It was early May, the mornings were still fresh, and I had been sleeping in the back of the van that at times felt more like a refrigerator than an abode! However, walking out in the morning dew with the sun just breaking over the horizon and the dawn chorus, music for those that can listen, was something I really looked forward to. I would*

*split kindling ready to fire up the rocket stove for my morning coffee and watch the carp in Bruce's pond begin their frisky little courtship dance, thrashing the margins into a foam. Learning to live closer to nature is certainly a far cry from the cities that I had once called home, but my journey into timber framing has given my well-being a significant boost!*

The volunteer days would start with a cup of tea or coffee and an explanation of what was due to happen. Mostly, people began as strangers to each other, but the work of the carpentry drew us together as companions. There would be a core group of about six people, with another ten coming when they could. It was always good to look up at the sound of the gate latch and see the return of a friendly face. Pete Cutforth always arrived on time on his electric bike. He would cut a graceful figure as he swooped down the curving track into the yard. Bruce Carley would always be there. Bruce had committed to put as much time as possible into building his new barn. Bruce's good-natured greetings and humour always set the tone for the rest of the day. Nothing is too much trouble for Bruce. He and his wife Esther grew up in the local area, so they had a wide circle of neighbours and old school friends who either came to help or to just call by for a cup of tea and a good look. The best time to visit was at tea break. Esther would arrive with a freshly baked cake, sometimes two – one for the morning, one for the afternoon. When she had time she would stay on to see what we had achieved or make an oak peg or two. Looking back on those days, Esther remembers:

*Learning some of the skills of timber framing, using the trees we felled locally and making them into something usable as well as beautiful was so intriguing. We loved meeting all the different workers and volunteers who joined us on the project and gained a lot of happiness from the camaraderie we shared.*

Bruce's father made weekly visits from the beginning, and I can still see the expression of disbelief as he looked at the piles of logs that had occupied the site for most of the winter. That look shifted to a look of genuine curiosity and then appreciative interest as the milling, marking and cutting progressed. By the early summer the frames were appearing and disappearing from the central stage of the framing yard, and there was a feeling that the barn was taking over with a kind of benign momentum of its own. Not quite building itself, but its character was becoming apparent. It had become a thing.

The lay-up for a wall frame is a big one. The entire length of the building has to be placed onto trestles. The walls always seem much larger in this position, and for that reason also very impressive. The length of the wall is such that one continuous piece of wood is not a possibility, so scarf joints are needed. We needed to make two joints for each top plate at the Carley Barn to achieve the total 13 metre length. There are numerous types of scarfs, and there is an interesting development of design over the centuries. The one we used is taken from the eighteenth-century Great Barn at Wallington. If there ever was a colloquial name, it has been lost. Now, it has a rather unglamorous technical name: a 'face-halved and bladed scarf with four edge pegs'. When the scarfs were ready, the plates were laid out, the jowl posts cut into them and then the rising braces that triangulate the whole wall frame.

The last timbers to be laid-up are the studs. These are secondary timbers, and their purpose is just to support the

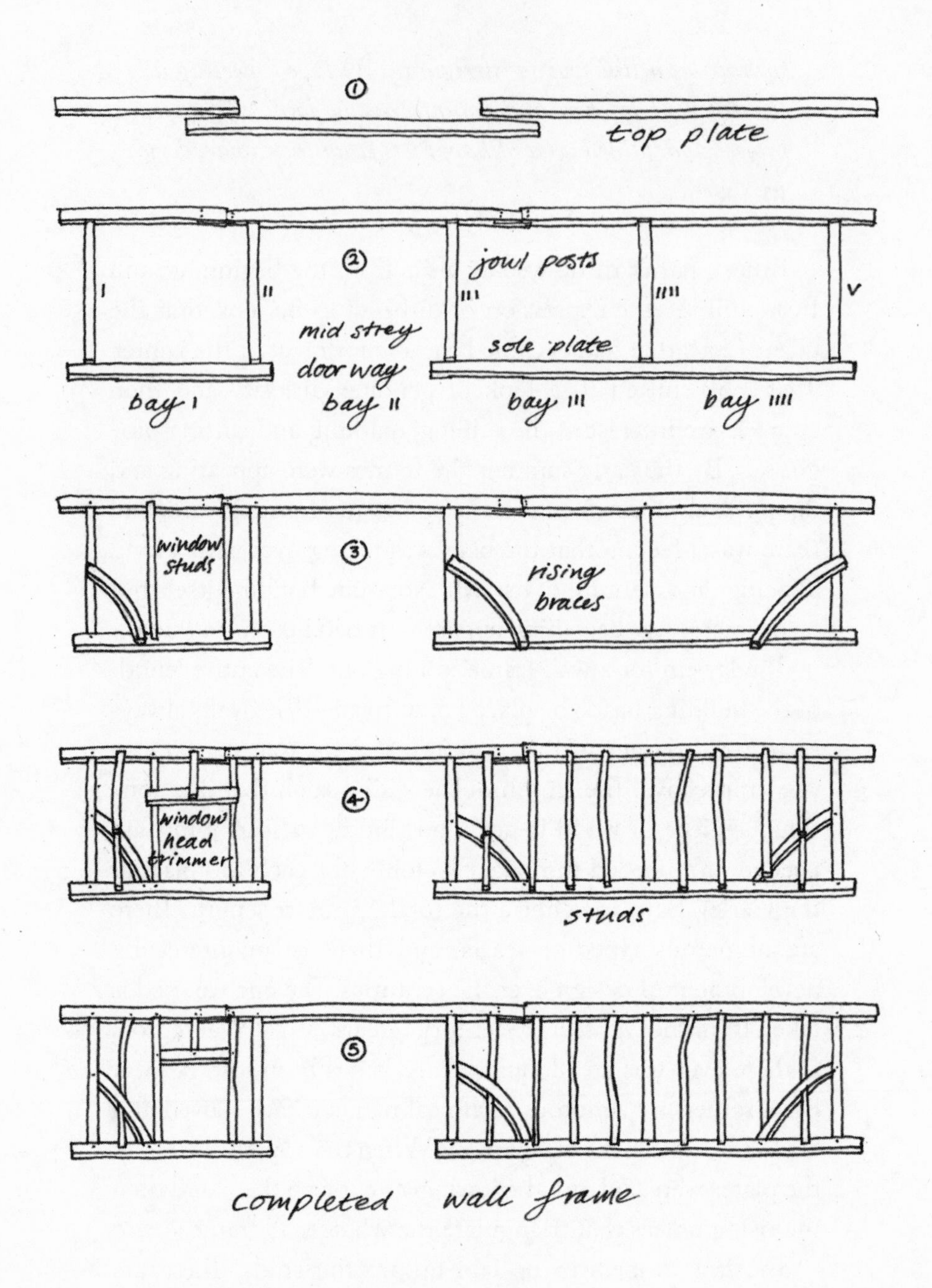


The sequence of lay-ups for the front wall frame.

external cladding. In old timber frames, they may be sawn from a variety of types of tree, not necessarily the same as the primary frame, and include more sapwood and sometimes less durable timber such as ash or even poplar.

Of all the timbers in the barn, studs are the most inconsistent. Most of them are re-sawn from the slab rails resulting from milling a primary timber like the principal rafters. Their irregularity of form, including the occasional curved timber, is counter-balanced by the regularity and rhythm of their placing at equal distances along each bay. To select studs, the volunteers would be armed with a drawknife, a plane, a clamp and a pile of various small-sectioned timbers and offcuts. With a would-be stud laid-up on two trestles, a little time and care is taken: drawing away the bark and bast, sawing off and planing any branch stubs and smoothing out any divots along their lengths. After this close inspection, rotten and split wood would be set aside and, with no other options available, these ultimate rejects found their way to Bruce's firewood pile.

With the studs temporarily in place, the wall frame is complete. At Churchfield Farm, the wall frame lay-ups coincided with some beautiful weather and sunshine. The wall frames were so long that they could easily accommodate a whole team of people, all chiselling away in a long line, making light work of all the various mortices and housings. These were happy days outside in the sunshine working together. Sometimes the levelling and aligning would take a bit of time. The marking-out would take a bit of head-scratching. So it wasn't all hard physical work. There was time to stop and take in the scenery whilst a problem of a particular scribed joint was resolved. There were memorable moments. As one volunteer and keen bird-watcher remembered:

*Sometimes the most wonderful natural scenes unfolded. One day, while working outside on one of the frames, I looked up to see a pair of red kites soaring and circling above the field. I was then astonished to see a pair of buzzards and a pair of kestrels had joined in. It's not every day you see that. It was fantastic.*

For the Barn Club members it was greatly heartening that other experienced timber framers wanted to get involved. High summer was upon us, and the date for the raising was booked. The pressure was on, so along with volunteers, the arrival of some experienced carpenters was a welcome boost to Barn Club. Rick Hartwell and Kris Vill set aside a week from their normal woodwork and pitched up to help. They brought with them the wonderful attitude that carpenters bring to a 'rendezvous', being happy to sleep in hammocks

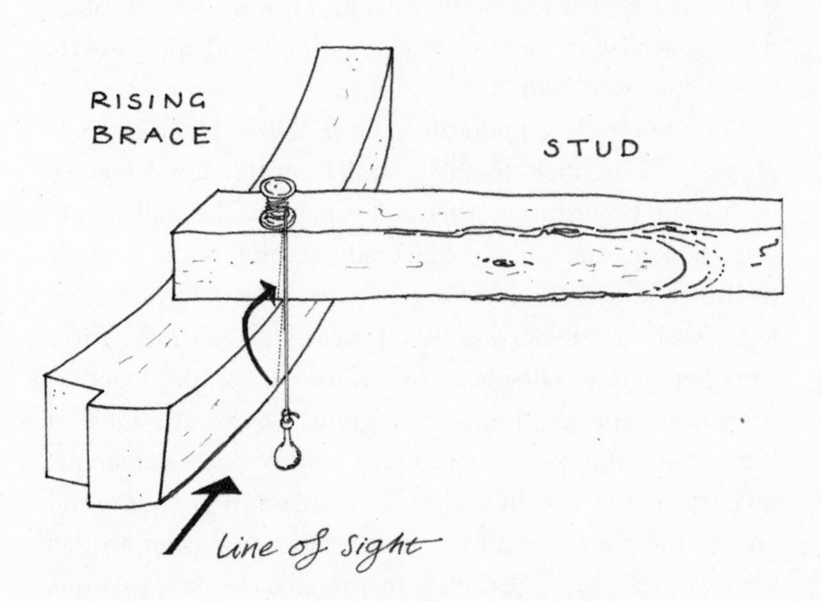


Scribing the shoulders of a stud to a rising brace.

strung between the arch braces of another barn nearby. Kris – wiry, energetic and without doubt the most talkative person on-site – commented to me afterwards:

> *The idea of having no power tools on-site was a novelty, but its impact was obvious from the start. Firstly, it made the project much more accessible. Limiting the construction to hand tools gave beginners the confidence to approach the challenge, to learn new skills and to be part of a communal creative effort. The workplace suddenly becomes much more sociable – less noise, less dust. It's possible to talk while you're stood in a line chopping out mortices on a summer's day, chatting comfortably while someone is using a handsaw nearby. People can get to know each other, questions can be asked, knowledge and experience shared, and a community formed. I hadn't realised how much I had missed working on a community project, with all the different people with different skills and interests you get to meet and talk with. It was a very interesting crowd that came together. I loved it.*

With the help of the more experienced carpenters, each day's achievements increased and the hours extended a little as well. When the weather is on your side, working late into the evening allows you to experience a mellow time of day that generates a kind of languid satisfaction. Working when the shadows are lengthening is very different from working under the glare of the sun overhead. If you can imagine a village cricket match extending into the gentle warmth and stillness of a long summer evening, with the smells of the meadows drawn out by the cooling air, you will get a sense of the good feelings that we felt as we chiselled and sawed the last of the carpentry joints after a long day.

## Tie Beam and Roof Lay-Ups

Cutting the dovetails on the tie beams and conducting a temporary roof raising at ground level were the final phases of the lay-ups. To check that all the purlins fitted and to scribe and cut the wind braces, it made sense to work in 3D rather than make a lay-up for each side of the roof. With the tie beams all settled into the dovetail housings in the top plate, all the components of each of the roof trusses are brought over from various cross-frame stacks and assembled one by one. The purlins and wind braces were scribed, cut and added to the structure in the sequence that they would eventually be raised.

At the top of the roof is a triangular ridge purlin. The ridge purlin rests on the small horizontal yokes at the apex of the cross-frame roofs. One of the beautiful and expressive parts of an old timber-framed building is the ridgeline. It may hump and dip in places, in contrast to modern roofs that are always straight. In historic buildings, the undulations may well be as a result of the roof timbers settling downwards over the centuries. But in fact, some ridgelines may have been curved from the start. Milling the triangular section for the ridge purlins accentuates the release of tensions in the timber, and they often spring into a subtle curve. The sections of the ridge then flow up and down from one bay into the other as they run along the roof. This flow will be visible when the ridge tiles are finally laid. One of the most welcome compliments about the completed barn came from a neighbour who lived close by. He walked across the valley one day to say how much he liked the shape of the roof and the colour of the reclaimed tiles. He said he loved 'the beautiful undulating roof, and how it seemed to glow when it caught the light of the rising sun in the morning.' Traditional barns always seem

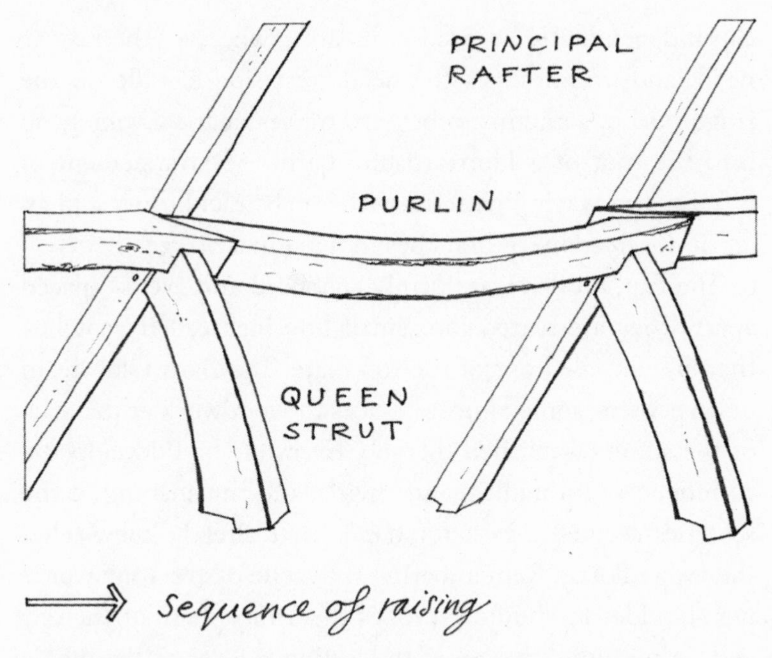


Bay-length purlins with their scarfs clasped by the queen struts.

to fit well in their landscapes, and it was good to hear that the Carley Barn did likewise.

The primary roof frame was finished, and all that remained was to make some preparations for the common rafters that would complete the roof structure. These rafters would be fitted one by one after the barn was raised. There were twenty common rafters on each side of the roof. These were milled to 4 inches deep from very young elm tree thinnings. Like the studs, they were sawn so that they are straight when viewed along just one face. This is the face that would become part of the roof plane for attaching battens and then tiles. Due to their often irregular curving shapes, the heads of each pair don't necessarily come into contact with each other. The triangular ridge purlin exists so that each rafter can independently be fixed at its apex, wherever the rafter's head lands. Sometimes the heads rest side by side on the ridge purlin, sometimes they are inches apart. Looking up into the roof of a Hertfordshire barn, the arrangement of rafters form waving patterns with an abstract beauty and an air of the freedom of branches moving in a breeze.

The feet, however, are firmly anchored and evenly spaced apart. They are seated into small housings called 'scotches' that are chiselled out of the top plate. The Barn Club team of carpenters, some almost novices to woodwork at the start of the summer, relished this job. Everyone had developed a confidence with mallets and chisels, and the shaping of the scotches seemed a task for hands that already knew what they were doing. Remembering the scene of everyone working shoulder to shoulder evokes the atmosphere of the day and, in my mind, is one of the lasting images of the whole project. Quick work was made of the forty scotches and soon the last one was done. The cutting of the roof was completed. With more than a little regret at having arrived at the end of the carpentry, the various components were dismantled and returned to their respective stacks.

Many months had been spent slowly developing each frame through the various lay-ups. It had been a long journey. Each piece of wood had been carefully examined, orienting them for bow, spring and wane. Each timber marked, scribed and cut; worked on to the accuracy of the thickness of a scratched line. Each piece of wood held, handled and understood for its potential and place in the frame. In total, 309 pieces of timber. After all that concerted effort it is good to sit down, wind down and reflect on all that has been achieved. I feel certain that the carpenters of the Great Barn at Wallington would have done the same. The sense of satisfaction was palpable. We all

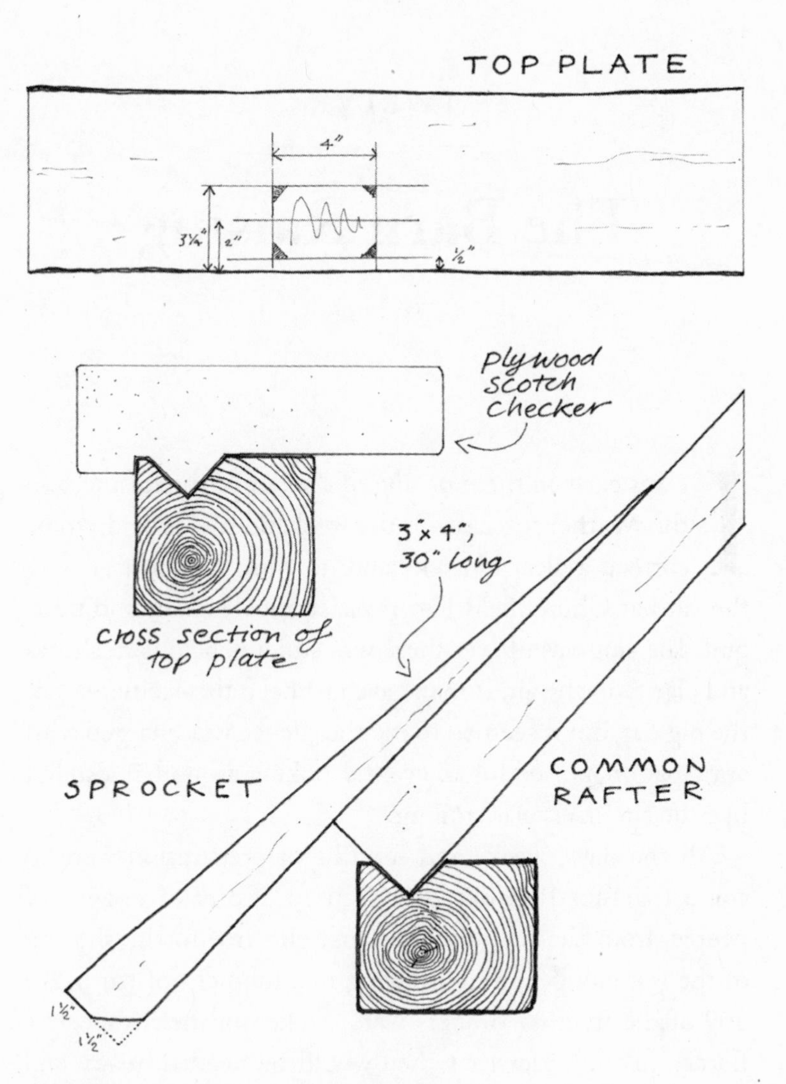


Scribing scotches to fit the rafter feet.

rightly felt proud of what must have amounted to thousands of hours of work between us. The memory of freezing winds and powder snow of the last day's milling earlier in the year had faded, and now we were looking ahead to the weekend in the diary when we would all gather again for the barn raising.

TWELVE

# The Barn Raising

It was early in the morning of the day of the raising, and the weather forecast for the weekend was proving to be correct: a clear blue sky and no wind. When I arrived, the yard at Churchfield Farm was strangely silent and tranquil. The sun was up, yet there was still a delicious freshness and clarity in the air. It may have just been the excitement of the big day, but it seemed to me the silence was charged with anticipation, poised for an event. Looking around, it also felt like the site itself was smiling.

All the days, weeks and months of preparation were to come together that day. The focused efforts of dozens of people, from the felling of the first elm tree to the shaving of the last oak peg, were drawing to a moment of truth. All 309 of the framing timbers were stacked in their respective frames in the order that they would be needed, ready and waiting. There was nothing left to do, and even if there had been, no time left to do it.

As usual, Bruce had been out even earlier to tend to the animals, fill the tea urn with water and switch it on. By the time anyone else arrived it would already be quietly simmering. As I started to unpack some tools from my truck, I saw that a fellow timber framer, Herbert, was also up and about. He seemed as excited as I was, and he remembered that morning well:

> *The whole raising weekend was an amazing event. I camped in my van overnight and woke up early the next morning, prepared my breakfast and sat down in a camp chair to watch the sunrise on what was going to be a beautiful day. At that moment I realised how lucky I was to be involved in such a great project and with such wonderful people. A community was about to arrive, to raise a new vernacular building into the landscape in front of me, for this generation and future ones to enjoy.*

Gradually, the Barn Club crew began to arrive and settle with a cup of tea, have a chat, and gaze across the site at the evidence of the work of their own hands. We all felt directly connected to the stacks of timber laid out around the framing yard in front of the brick plinth of the barn. These stacks of jowl posts, curving braces, long plates and rafters and endless studs. All neat, deliberately ordered, and each piece with its unique identifying carpenter's mark. The tenon ends protruding from a stack of jowl posts reminded me of the piles of salvaged ancient timber from the barn I had helped demolish in Sussex many summers before. In the intervening years I had gathered the experience to know what all the different timbers were for and how this enormous jigsaw was to be completed.

A frame raising entirely by hand is a very rare event these days. It doesn't need to be, but machines, particularly big strong ones, have replaced human muscle. The knowledge of how to do it, or even that it can be done at all, has gone from our rural communities. It's good to have the machines to make life easier, but a shame to lose the skills and confidence that lets us do these things for ourselves. It felt very empowering to know that the several tons of wood in the yard that morning could all be lifted and placed in position by our own muscle power. From previous experience, I also knew that it wouldn't be too arduous. With enough people, anything is possible. Though, obviously, lifting heavy objects above head height can be a risky business, so a hand raising needs to be approached thoughtfully and with great care. A frame raising is a highly organised event.

The day before the raising the sole plate was assembled on the plinth and everything had been double-checked. It must be the same for many 'big day' events. You run through all sorts of scenarios in your head and worry about things to find the weakest link. You walk around trying to picture what was about to happen and see if anything is missing. My eyes landed on the peg buckets. The ends of some pegs were looking a bit ropey. So I checked through them and sorted out the slight seconds, and one or two that were best described as 'curiosity pegs'. Some were just a little short or too small at the head. These 'shorties' would be fine for the studwork, but not the primary frame. Those that were cigar-shaped could be slimmed down to the desired gradually diminishing shape with a drawknife. A block with a ¾-inch test hole is a useful peg-checker. If the peg slips easily into the hole until two-thirds of the way up, it will be okay. If the peg falls straight through, it goes into the kindling bucket. A last-minute peg-shaving session brought the total up to the 303 pegs required. With that done, there was nothing left to do but go home and rest. . . as if that were possible amidst all the excitement!

In the past, whole communities gathered to raise buildings, just as they did for harvests and other tasks. This principle, of everyone joining in together, is part of the joy of the day. A frame raising by hand amounts to a festival, with food and drink provided. For days in advance, Esther Carley had been preparing the food with help from local friends. A long list of people had been contacted to let them know about the raising. There were old friends, new friends, relatives and neighbours from the village, and of course all the volunteers and their own guests who were curious to see what Barn Club was all about. Whilst we waited for enough people to swing the first frame into position, Herbert and a small team secured short props to prevent the sole plates from being pushed sideways as the frames swung upwards.

As people arrived, I encouraged them to feel they could become involved, and join the raising team, even if for a short while – and many did just that in whatever way they felt comfortable with. Some had no previous knowledge of woodwork, let alone timber framing, and yet were more than capable of pulling on a rope when given a signal. Some had a lot of experience. None more so than timber framer Joel Hendry, who came up from Devon to co-ordinate events along with his teenage son Marlin. There were another seven framing carpenters as well, who would be familiar with the components of a frame. With the volunteers included, and others who came just for the day, there were forty-eight people eager and willing to help. As the 10am start time approached, the site was filling with a throng of people, and excitement was building.

The church clock chimed and the tenth stroke rang across the meadow. It was time to begin. After a quick head-count we knew that enough people had arrived to begin the first lift. The white rope was drawn across to separate the raising crew from spectators and late arrivals. Blue hardhats were passed around. There were to be many stages to the raising with breaks in between, so there would be time and opportunity for people to drop out to rest or others to join in.

After a few words of welcome from me, and explanation of the day ahead, Joel took over and set the scene for how things would take place and what was about to happen. Before each lift, there would be a short coaching session like this so that everyone would know what to do and how to do it – and how to do it safely. We had two people on-site with up-to-date first aid certificates, and they were identified to everyone. Marlin was given the title of 'Peg Meister', meaning he would hand out the pegs, matching the length and quality of peg for each mortice and tenon joint.

The first frame to be raised was the back wall frame. But first it had to be pegged together lying horizontally. The timbers in their stacks looked very heavy, but after many months of carpentry, the volunteers knew how to lift and move them around without endangering anyone's back. By now they knew to bend their knees before lifting, to keep their backs as upright as possible and to hold the timber close to their bodies rather than leaning out over the load. Trestles were positioned along the barn's length and the top plate timbers laid on top. Then the scarfs were lined up and brought together with the help of a heavy mallet. Oak pegs, knocked home and tight, brought the top plate pieces together to act as one long length.

The jowl posts were next on the scene, and the tenons at their heads were fed into the mortices in the top plate. Their feet were placed on the sole plate inline with its mortices. Pegs were again allocated and knocked in, always from the face side of the frame. Rising braces also were brought from their stack and each carpenter's mark matched to the right

A pikestaff made from a spruce tree sapling for lifting the wall frame upright.

jowl post. Teepins were used to hold the braces to the jowl posts. These were used so that they could be loosened to help align the brace feet to the mortices in the sole plate. Long poles called pikestaffs were also arranged along the frame, heads resting on the top plate and feet towards the front wall. The ropes were looped over the top plate and wound round and round, down the lengths of the pikestaffs. The tenons of the jowl post feet would be guided to their corresponding sole plate mortices as the frame was raised, using 4-foot-long scaffolding tubes with flattened ends.

With so many people wanting to help, we decided to line the entire length of the top plate with people standing shoulder to shoulder. Others stood back holding the pike-staffs. Each jowl post foot had a person ready to guide it into its place. Everyone had a specific job and stood ready and poised, and a little bit tense.

The first task was to simply lift the top plate an inch or two, hold it there, then lower it again. This was to get a feel of the weight and check the most comfortable position to stand. The talking ceased, and all eyes and ears were on Joel. On the count of three the top plate was momentarily raised and then lowered back onto the trestles. There was unified sigh of relief. The frame wasn't nearly as heavy as everyone had been expecting, which was good, and the task ahead began to seem possible.

The next lift was for real, and on Joel's command the frame swung skywards. When the top plate was at shoulder height, the pikestaffs took the strain and as the frame became more upright the jowl post tenons began to slide into their mortices. Literally in a matter of seconds the whole frame had been raised and was in place. The pikestaff holders steadied the frame in a plumb vertical position without any effort. A round of applause broke out and everyone relaxed. It had been an intense few minutes. When the raising happens, it happens fast!

Before moving on, two rising braces that run at right angles into the corner jowl posts had to be fitted. Using the pikestaffs to steady the structure, the frame was leant over slightly so that the brace tenons could slip into position.

Additional lengths of 4″ × 2″ scantling were screwed front and back of every other jowl post as temporary props. It all looked very solid and strong, as indeed it was.

The first frame was up, and before switching to the next one, and to allow our adrenaline levels to settle, we stopped for tea and a slice of cake. Cakes, generous supplies of cakes, had been a theme throughout the months of carpentry. In all weathers Esther and a close family friend of the Carleys, Diane Sandbrook, would arrive with just the right refreshment needed for hungry volunteers. The day of the raising was no exception.

After a good break and a lot of talking as more people arrived, the raising crew gathered once more in the yard and donned blue helmets. A few people swopped positions with newcomers, to share out the experience. Next up: the front wall frame. This lift was talked through and explained by Joel as before. Props were moved from the back of the barn to new positions to brace the front wall sole plates. Trestles were set up in a line in front of the front wall plinth. The top plate members were wheeled across on the timber trolley and lifted onto them as before, and the scarf joints assembled and pegged. The jowl posts, rising braces and window frame studs were also all assembled. This time the frame was face-down, ready to swing up facing outwards. Lastly, pikestaffs were roped into position.

With everything and everybody ready, the same procedure as the back wall frame was followed. With a 'three, two, one, lift' from Joel, the frame swung upwards in one steady, easy movement. This time the rising braces at each gable end of the building were held up ready to meet the frame rearing up towards them. As the frame neared a vertical position, the tenons of these braces were fed into the mortices in the jowl posts. This was another textbook lift that seemed to achieve far more than the physical effort taken.

With a mixture of relief and wonder, everyone stood back to see what had been achieved. The timbers were in exactly the same relative positions as they had been in a horizontal plane during the lay-ups. But now that they were upright, they appeared bigger, and the frame as a whole expressed a much stronger presence. This impression grew and grew as the raising progressed. The cross-frames were next.

Unlike wall frames, each of the cross-frames has a tie beam in place of a top plate. Rather than raise the cross-frames in one piece from the ground, they are assembled piecemeal, with the tie beam acting as a cap on all the upwards-pointing timbers in each frame's face. The tie beams dovetail joints also lock across the wall frame top plates by being lowered from above. So the next phase in the sequence of raising was to lift these heavy beams up above the level of the top plates. With a block and tackle, a pair of 'sheerlegs' and enough people pulling on a rope, this can be managed without a crane.

A block and tackle is an ingenious and simple way of multiplying the strength of people pulling on a rope. A block is a casing with one or more pulley wheels within it. The tackle is the term for the hooks and strops. Usually there are two blocks. A rope is tied to a ring on one block, threaded around the pulley wheel of the other block, back round the pulley wheel of the first and so on until there are no more wheels left to thread the rope round. When the loose end of the rope is pulled, the two blocks move towards each other. The sum of the number of times the rope goes back and forth between the blocks is the multiplier for the strength of the people pulling the rope. Some pulling power is lost through friction of the bearings of the pulley wheels, but the multiplication of strength is staggering to experience.

The sheerlegs were a sort of 'A' frame that had its feet securely positioned on the ground and an upwards-pointing arm to which the top block of the block and tackle was fixed. There was also a strong line running back from the top, anchored securely so that the sheerlegs could not topple over. We attached a double-block at the head of the sheerlegs, which means it had two wheels within it, and a single block with a lifting hook hanging vertically beneath it. The load would be attached to this lower hook. We assumed a four-fold multiplier for the pulling strength of the people on the loose end of the rope. Using the block and tackle, a beam weighing a quarter of a ton could easily be lifted by four people. We added four more volunteers as a margin of safety in case anyone slipped as they were pulling the rope. In the event, due to the strength of the people wanting to join in, we probably had three times the pulling power that we needed.

Beginning with the gable cross-frame furthest away from the anchor point, cross-frame-V, the sheerlegs were set up and the tie beam wheeled underneath. Two scaffold towers were erected outside the frame at either end of where the tie beam was to go. The scaffold decking was placed at the right height for people to comfortably guide the tie beam into position. Two ropes were attached to each foot of the sheerlegs, running out of the building and securely anchored to prevent the legs slipping as the beam was lifted. By drawing in the back anchor line with a cable winch, the inclination of the sheerlegs was adjusted to about 75 degrees from horizontal. A webbing sling was wrapped several times around the tie beam at its middle and looped onto the hook of the lower block. With pegs at the ready, everything was set to start the lift.

On Joel's command the rope team pulled and the beam gradually lifted from the ground and began its ascent. As with the lifting of the wall frames, its progress seemed effortless. When the tie beam had reached a height just above the top plates, the lever on the cable winch was pumped to and fro by Gordon and Marlin, the most senior and junior of the raising team. As the sheerlegs' head rose more vertical, the tie beam inched across until mortices were in line with the jowl posts' protruding tenons, its 'teazle' tenons. The rope-pulling team slowly walked the rope back towards the sheerlegs, allowing the tie beam to lower over the jowl posts and mid-posts. Finally, the dovetails dropped into their housings in the top plates. All the joints fitted perfectly, and with oak pegs knocked in to replace the teepins, the first of the cross-frames was in place. A good moment to stop for lunch.

The food was delicious and there was a real buzz of excitement and laughter as people chatted and shared their impressions and memories of the previous months of carpentry work. After the last cup of tea or coffee was drained, there was also an eagerness to crack on with the raising. A sizable team regrouped in the framing yard, put on their blue hats, and were ready for Joel to explain the process again for the benefit of those who had just arrived.

The afternoon's work began, and the tie beams went up one by one. Each time the sheerlegs were moved further along, the winch took up the slack and the next tie beam was lifted. The open cross-frames had arched braces to align as the tie beam was lowered. These are best left fairly loose in their jowl post mortices until the tie beam is finally in place; then the pegs can be knocked home good and tight.

Other cross-frames had more tenon ends of posts and studs that needed to be fed into place. The occasional teazle tenon refused to slip as comfortably as intended. At this point a giant wooden mallet was passed upwards and with a substantial whack, Joel 'encouraged' the tenon into place.

To give the crew of thirty or so people something to do whilst the sheerlegs were being repositioned, all the secondary timbers, the studs for the walls, were fitted. Each stud had a number chiselled or gouged into its head, and had a matching number somewhere on the frame. A stud-frenzy ensued. Everyone rushed around with their chosen stud in their hands looking for the right place for it, like a giant game of pin-the-tail-on-the-donkey. Somehow or other, each stud found its right place and was swung into position, pegged at its head and nailed through the beard at its foot. The walls now looked like they were enclosing real spaces, but there was too much to do to stop and admire the work. Time for that later on.

Since the roof had previously been raised at plinth level, we were confident that the purlins and wind braces would fit. This time it would be done at height using three scaffold towers. If anything, it was easier than before because the timbers were all fettled, ready to fit. However, working at height requires more awareness of risks and dangers. Apart from physically falling yourself, objects can fall on anyone below. So, whilst the roof was assembled, everything was handed up from outside the building, which meant that no one needed to stand immediately underneath the work area.

The roof truss for cross-frame-V went up first, this time with its small-section gable studs added as well. It was temporarily propped whilst the towers were moved along for the next truss. Once the roof truss for cross-frame-IIII had also gone up, the bay-length purlins and ridge were inserted, and then the wind braces were added. The roof structure then became self-supporting and stable, ready for the next bay. To lock the purlins' scarfs together, holes were augered through the backs of the principal rafters, through the middle of the purlin splays and into the head of the queen struts. Rather than the standard oak pegs, square ones were used. This is because the hole was straight, without an offset or drawbore. Instead, the edges of the square pegs would cut into the sides of the round holes and lock tight. The square peg would work like a nail. But they have to be just the right size. A square peg 1 millimetre too wide jams and will buckle and snap rather than go into its round hole.

As the roof team progressed, the barn began to look like a real building with a proper roof-shaped roof on top. It went like clockwork. The system of scribing and cutting the roof truss joints on a horizontal plane, then assembling the same timbers vertically worked without fault. But just as ingenious is the principle of cutting slop into the throat and front cheek of every mortice, allowing us to assemble the joints in a fluid motion and providing enough wriggle room for a number of timbers to be aligned at once before closing the shoulders tight. With the shoulders cut so carefully to match each bevel, the structures were surprisingly stable, even before pegs were inserted. By the time they were all knocked home, the whole frame became as tight as a drum. The combination of a peg's wedge shape and the offset of the drawbore pulled the nose and back face of each tenon into firm contact with its corresponding mortice cheeks, and in doing so brought the whole structure into alignment. In those millimetre-accurate positions, the shoulders all locked tight as well. It's like those children's wooden figures that go floppy when your thumb pushes up underneath, then spring back tight when you let go.

The closing moments of the raising were fast approaching, and a crowd of onlookers gathered in the yard below to watch the final scene. The last yoke was nailed into place at the head of the last rafter pair of cross-frame-I, ready to take the ridge purlin: the final piece of wood for the raising. I watched as it made its way across the yard to be handed up to Bruce, who was with Joel on the top scaffold tower. The time had nearly come to celebrate the topping-out.

It's good to celebrate an achievement. Creating and raising a barn doesn't happen every day, so the moment that the last piece of the roof goes up calls for a bit of a show. There is a tradition of fixing a living branch to the top of a roof when this moment arrives. Topping-out, as it's called, probably goes back to our Celtic forebears who knew the nature spirits that live in trees. The exact meaning of topping-out is lost, but as a tradition it has been handed down through the centuries. I feel certain that the carpenters at the raising of the Great Barn at Wallington would have nominated someone to climb to the top of the last roof truss to nail a sprig from a living tree to the apex. Perhaps followed by a libation? Who knows, but the opportunity is there to make the moment special and mark the occasion in a dramatic way.

I asked Bruce to wait a minute before he climbed back down and I found Esther to see if she would say a few words. Both of them agreed. Living elm would obviously be appropriate for this barn, and as it happened there were several elm trees growing in the hedges of the meadow. I went to cut a sprig from one whilst Esther thought about what she would say. Meanwhile, everyone organised themselves into a long line snaking across the yard from the gable of the barn.

The sprig of elm for the topping-out passed along the chain to Bruce.

I found a healthy stem of elm leaves and brought it back to Esther and handed it to her. She was at one end of the human chain and her husband Bruce was high up on the roof of the barn at the other.

At that point in the raising we had all become part of something bigger than any of us. The barn, the elm trees, the carpenters, the volunteers and the onlookers: we were now all acting together in a scene from a tradition that was coming back to life. Everyone hushed and Esther said words that completely fitted the occasion. She thanked everyone for what they had done, in whatever way, and welcomed the people who, over time, would visit the barn in the future and the inspirations she hoped for them. It was an acknowledgement that every effort counts, and that everyone matters and is valued, including people who feel they are 'square pegs in round holes'. The elm branch was passed along the line and each person passed their thanks, joy and good wishes along with it. It made its way into the barn and up the platforms of the tower to Bruce. Bruce held it aloft and everyone cheered as he nailed it to the apex of the roof. I have never seen such a big smile as the one on Bruce's face at that moment. We had done it!

After that, what could we do? We stopped for tea.

In fact, nobody really wanted to stop for long. After a short break, everyone was inspired to carry on, and some of those who had just come to watch wanted to join in. So they did. There was an all-encompassing willingness and desire to help. All of the pegs needed to be checked and missing ones knocked-in and tops and tails sawn off to leave just ¾ inch sticking out. There were twelve floor joists to prepare, which involved an impromptu lesson for newcomers on marking and cutting shoulders and tenons. The ends of the floor joists that weren't jointed to the central floor beam would rest on two rails fixed on the inside of the wall studs. These rails had to be levelled, drilled and spiked. The upper gable studs were to be to fitted and sole plate horns marked and sawn off.

There were enough jobs to keep everyone busy for an hour or so. But forty willing hands can get a lot done quickly. The towers were dismantled and the temporary props removed and tidied away. Brooms were produced and the offcuts were swept clear. Just as the afternoon began to slip into evening, we reached the point when there was nothing left to do. The focus and energy that had propelled me through the day had found its finish line. It was done. The barn was raised. I needed to sit down!

It's difficult to describe my feelings as I let the achievement of the day seep in. I guess I felt a kind of rapture. I wondered: did they also feel like this? The village carpenters who built George Orwell's barn in 1786? Did they feel what I was feeling at that moment? It wasn't just the joy of seeing the frame upright and all the timbers in place, nor the relief that we had met the challenge of doing it safely. It was more like an overwhelming feeling of amazement and humility due to the enormous scale of the erected frame that now dwarfed us. I wasn't the only one, either. Those feelings were in the eyes of other members of Barn Club as well. In the following half an hour we wandered around, fully experiencing the frame in three dimensions for the first time. We had commanded the wood for so many months, wrought it with our sharp tools and transformed the tree shapes into multiple jointed timbers. Now these same timbers arced and towered overhead and we were once again gazing upwards at their splendour just as we had done looking up in awe at the mighty trees in Bushy Leys Spring wood nearly a year before. This time, the exposed honey-coloured elm seemed to glow against the azure blue of the sky and everything seemed to be smiling.

Sitting down on the sole plate, supported by studs either side, I could look around and for the first time take it all in. Being inside the three-dimensional frame was very different from seeing everything in two dimensions as lay-ups. There were interesting vistas to absorb and appreciate, past jowl posts and arching braces to the roof frames beyond. And the old oak tree in the neighbouring field to the west, perfectly framed by the mid-strey doorway. For the first time we could truly see the building that we had spent so many hours creating. Some timbers I recognised for their particular waney edges, finally seeing them in the context of the surrounding frame. Other pieces of wood brought back memories of watching a volunteer spend hours shaping and cherishing the flow of its grain. I looked around, allowing the barn to speak to me, its beauty finally revealed and radiant. Why it seemed so beautiful, I couldn't say. But it was so much more than just a barn.

Now that it was up, we intended to celebrate. Esther led the team to set out a feast within the main barn space. Trestles and scaffold boards were converted to long tables, tablecloths and flowers made them ready for a banquet. A horse trough was found and filled with water to keep the drink bottles and cans cool. Lights and lanterns were strung around the beams and braces. It looked fantastic as the sun began to dip, sending a warm glow straight through the mid-strey.

As if to complete the sense of a Hollywood film script being played out, as we were about to fill our plates with food, Susannah, the vicar from the church across the meadow, arrived and offered to bless the barn. She had been in the church preparing for a service the following day and noticed the sudden appearance of the timber frame. Her timing was perfect. In a fitting counterpoint to the more pagan topping-out ceremony just hours beforehand, she wetted another branch of elm in the drinks-cooling horse trough and used it to spray water, as she improvised a blessing on several of the jowl posts nearby, suppressing her own laughter as she did so. It was a bit surreal, strangely appropriate and perfectly timed for the beginning of the meal, which of course Susannah joined us for.

The day had ended in a way that village carpenters and communities of the past would recognise. A fire was lit in the yard as the darkness came on. There were lots of warm embraces and expressions of joy at what had been achieved. The children chased around the barn, climbing in and out of the studs, and some of the youngsters dared each other to swim across the pond. That day we had raised a barn in the village and had brought to life a tradition that seems so human, generous and natural. We had made it out of elm trees from a local wood, milled it on-site and cut the joints right in front of where it was to stand. Through Barn Club, we had made the whole process open and inclusive to anyone who was interested. Lastly, with the combined strength of many willing arms, and with joy, we had lifted the timber frame high into position. It was worth celebrating!

We drank and ate and laughed well into the night. I can honestly say that I can't actually remember going home at all.

# Conclusion

The barn now stands completed and looks comfortably at home in the meadow at Churchfield Farm. It looks as if it has always been there, or was always meant to be there. Passersby will see it and appreciate it for many years to come. But what they will not see are the people who made it, the struggles to scribe the joints correctly, or the laughter, the stories and companionship. Just like the Great Barn at Wallington, they may see the marks on the wood, but they won't see the tools that made them. Nor will they have the memories of the elm trees as they towered overhead back in Bushy Leys Spring wood. In years to come, someone in Tewin will turn to their grandchildren and say, 'See that barn? I helped raise that. We did it by hand. I remember that day.'

Above all else, the vision of following the footsteps of village carpenters that we brought to life in the making and raising of the Carley Barn, has shown that a craft can be something that is lived and shared with other people. It is a process of bringing something substantial into being, together, whereby the use of hand tools nurtures all sorts of positive possibilities. With village carpentry the workmanship is there for all to see, but it is the healthy local woodlands, the trees and the people that count just as much. For the carpenters in Wallington over two hundred years ago, there may have been few other options. For us now, in the early twenty-first century, we need to make choices. We could consciously decide to restore our landscapes to health so that they can provide the timber we use, and to deliberately set out to create opportunities for people to gather together out of doors and be productive in meaningful ways. It is all there for us to choose to do. It's up to us.

After the raising of the Carley Barn, the volunteers were completely inspired and eager to carry on and help with the completion of the building. Brian expressed the sentiment felt by all the regular volunteers when he stated, 'Once I'd started I just had to see it through to the finish.' In the ensuing months we shaped and fitted the elm rafters, tiled the roof and clad the walls with weatherboard. The final tasks were to fit the internal joinery: the stairs to the storage loft, the oak floor and the doors. One of those was a simple planked door and had a traditional wooden finger latch made for it. Jo took on the job of making the handle on the final day of Barn Club. He has the last word:

> *A simple but nonetheless valuable thing for me personally was shaping a door handle. Why should a small object like this hold such sway in my mind? I can't really say, except that it is a work-a-day thing that people will probably think little of, but which gives both access and closure. It's the handle on the inner door leading from the lower room into the main barn. It doesn't turn, it just provides the means of taking a firm hold of the door; but I made it from the oak we milled ourselves and then dried. It has a lovely feel, and every time I visit in the future I know it is a little part of me that will never leave.*

# Glossary of Terms

Like most crafts, the words that are used have been handed down by word of mouth over the centuries to communicate parts of the materials or processes involved, un-noted by literature. The words of village carpenters were heard and not seen. Some have regional expression, like 'podger'; other words have inconsistent meanings and spellings, depending on where you are. Some words have become regulated, technical and universal. They appear in Wikipedia. They are like domesticated livestock, whilst others are still wild and have yet to be ear-tagged. Without doubt, many words have been lost. But new words arrive. Riggle: a neat trick to solve a carpentry problem, particularly in setting out and marking. This is the first time I have ever seen it written down. You may not agree with some of the following definitions, but accept them as quirks of evolution rather than pedigree animals. Generally, I have avoided words that people don't normally use, so the following should be broadly understood by carpenters everywhere. Although they are also descriptions of things that could be equally achieved with out words at all.

**Abutment** An intersection of two pieces of wood.

**Adjustable square** An L-shaped tool with one adjustable arm.

**Annual ring** A visible continuous loop in the cross-section of a tree, marking the previous twelve months of growth.

**Arch brace** A curving structural timber between a vertical post and a horizontal timber.

**Arris** The corner edge of a timber.

**Auger** A drilling tool shaped like a corkscrew, for drilling a hole.

**Auger bit** An interchangeable auger that can be attached to a drill.

**Awl** A small tool with a spike used to make a hole in a material, or if the point is sharpened, to scratch accurate marks on the surface of timber.

**Back-cheek** The bottom surface of the square mortice box furthest from the face of the timber.

**Band saw** A saw blade formed into a continuous flexible band.

**Bare-faced tenon** A tenon with a shoulder on one side only.

**Bast** The layer of vascular tissues immediately inside the bark of a tree that transports sugars and nutrients from the leaves to all other parts of the tree. Also known as phloem.

**Baulk or Balk** A length of timber with four or more sides that follows the original shape of the tree, its sides having been hewn using axes to remove the sapwood.

**Bay-corner mark** One of the four points that mark the intersection of the theoretical frame faces.

**Beam** A load-bearing horizontal timber.

**Beard** A projecting wedge-shape.

**Beetle** A large, heavy wooden mallet.

**Bevel** A sloping surface.

**Blade** The inwards-facing surface of a scarf on one timber overlapping a similar surface on a second when joining two lengths of timber together end to end.

**Blank** A piece of wood ready to be shaped.

**Block and tackle** Lifting equipment consisting of rope, hooks and pulley wheels.

**Bow** An upward bending curve in a length of wood.

**Bole** The base of a tree, including where the stem continues below ground before dividing into roots.

**Box-hearted timber** A length of wood with the centre line of the tree from which it was sawn running inside it.

**Bridle** The end of a piece of wood shaped like a tuning fork.

**Bridle joint** The connection between two pieces of wood, one end with a bridle and the other with a tongue of wood that fits into the bridle.

**Bruzz** A long chisel V- or square-shaped in section for cutting mortice corners.

**Cambium** A layer of living tissue encompassing all parts of the tree just under the bark and at the edge of the sapwood.

**Canopy** The top layer of leaves and branches in a woodland.

**Cant** The initial four-sided shape cut by a sawmill from a log.

**Cant hook** A simple leaver with an attached hook for rolling logs ready for milling.

**Carpenters square** A solid metal right angle measuring 2 feet by 2 inches on one arm, and 16 inches by 1½ inches on the other arm.

**Cellulose** The organic chemical compound giving strength and flexibility to the fibres in wood.

**Chamfer** The flat surface cut at 45 degrees along the edge between two perpendicular faces.

**Chin** A step in the surface of a timber, for example the base on a jowl post.

**Chisel** A flat metal tool sharpened at the end for cutting wood, often with a wooden handle and hit with a mallet. Chisels that are curved in cross-section are called gouges.

**Chord** The upper or lower surface of a rafter or other timber: the top-chord and bottom-chord.

**Circular saw** A metal disc with saw teeth around its perimeter and powered through a central axle.

**Cladding** The outer-most weatherproof layer of wood on a barn wall.

**Cleaving** Splitting a timber along its length in the direction of the wood fibres.

**Cleft** Past tense of cleaving.

**Collar** Horizontal timber spanning between two opposing rafters, usually under compression and approximately halfway or further up the rafter length.

**Common rafters** The sloping timbers that are laid on a roof to support the roof covering, usually supported by a purlin at their mid-length.

**Compasses** See dividers.

**Coppicing** Traditional woodland management system of periodically felling trees at ground level, to provide timber poles and resulting in vigorous replacement regrowth. Not all tree species will re-sprout in this way.

**Corbelled** A projecting lip of masonry that adds width to a wall.

**Cross-grain** A weakness in a length of timber when the grain runs at a slant to the edge of the wood rather than in line with it. Caused when the shape of the milled timber doesn't match the shape of the tree.

**Cross-cut saw** A saw with its teeth designed to cut across the fibres of a piece of wood.

**Cross-frame** The flat frame that spans between front and back walls, usually at right angles.

**Crown-cut** The position of the mill blade near to the top of a log that produces ripple and arch shapes in the grain of the timber surface.

**Crown post** A post in a roof truss rising from a tie beam and supporting a central longitudinal timber, or plate, that in turn supports the collars above it.

**Cup shake** A crack in a timber that follows the circular shape of an annual ring.

**Cutting list** The list of timber sizes, lengths and quantities required, as measured from the framing drawings.

**Dividers** A pair of metal arms hinged where they join at one end, with sharpened points at the other. The arms can be swung open and adjusted to set the points at a particular distance apart.

**Doated** Patches or speckles on the surface of timber that is white or lighter in colour due to fungal attack.

**Double-cutting** The second marking of the shoulders for the tenon timber that has been already partially inserted into its corresponding mortice.

**Dovetail** A type of tying or locking joint with a fan-shaped piece of wood fitting into a similar-shaped housing.

**Drawbore** Offsetting the holes for the pegs in a mortice and tenon joint.

**Drawknife** A long blade with wooden handles at each end to pull towards you and slice shavings from the surface of a piece of wood.

**Drop** The distance that a horizontal timber that rests on another is to be lowered by so that the surfaces align.

**Dutch elm disease** A type of fungal infection of elm trees *Ophiostoma novo-ulmi* that arrived in the UK approximately fifty years ago and killed nearly all native elm trees. Discovered by Dutch scientists and spread by elm bark scolytus beetles.

**Ear** A projecting end of a timber left to over-sail the joint, to be left intact to retain strength or sawn away on completion.

**Early wood** The first part of growth of each annual ring that has large or densely grouped vessels for transporting sap to the leaves in early spring.

**Eaves** The lowest part of an overhanging sloping roof.

**Face** The side of a timber facing outwards on the plane of a frame.

**Face-edge** A nominated edge adjacent to the face of a length of timber.

**Fell** To cut a tree down.

**Felloes** The curved outer parts of a wagon wheel.

**Fettling** Using a cross-cut saw to cut a kerf around all the shoulders of a tenon timber of a tightly assembled joint.

**Flag** A chiselled additional short arm on a carpenters' mark to indicate a particular timber's position within either the front or rear of a timber-framed structure.

**Flange** A projecting strip or lip of wood.

**Flat frames** The principle of each frame being marked and cut so that the members come together on a continuous flat plane.

**Foot** The bottom or lowest part of a timber.

**Foxy** Orange fungal colouring of timber due to brown rot.

**Frame** A collection of timbers aligned into a single plane.

**Framing drawings** Drawings of each flat frame, to scale and with key measurements, with the face of the frame facing the viewer.

**Froe** A rugged metal tool used to split wood when struck with a maul, consisting of a long blade with a handle at right angles at one end.

**Front-cheek** The upper surface of the square mortice-box closest to the face of the timber.

**Gable** The vertical wall and roof shape at the end of a building.

**Gable cross-frame** The plane of timbers forming a gable.

**Gauge block** The part of the mortice gauge that slides along an edge.

**Ghost-line** An initial marking line, often a dashed pencil line, that has the correct slope but not in the finalised position.

**Glade** A clearing in a wood.

**Gouge** A chisel with a curved cross-section.

**Grain** The direction of growth of fibres and vessels along the length of a tree stem or branch, indicated by the longitudinal lines of annual growth in the cut stem.

**Green wood** Freshly sawn timber with its vessels and cells full of water.

**Gusset** A flat material, such as plywood, that laps, and is fixed across a joint and strengthens it.

**Hammer beam** A short, cantilevered horizontal timber at the foot of a roof truss projecting inwards from a plate, supported by an arch brace beneath it and supporting a post above.

**Hardwood** Species of trees that lose their leaves in winter. Also called broadleaved trees.

**Haunch** The part of the tongue of a tenon that is removed so that it doesn't span the full width of the timber. The mortice is correspondingly also reduced by the extent of the haunch.

**Head** The end of a timber that will be at the top when the frame is raised.

**Heart shake** The cracks from the middle of a tree that radiate out when the ends are cut and the timber starts to dry.

**Heartwood** The non-living wood surrounding the middle of a tree stem and beneath the perimeter sapwood. It is generally darker than sapwood, the vessels are usually blocked and infused with chemicals that preserve them and make heartwood more durable than sapwood.

**Hedge row** An outdated term describing a hedge widened by several metres along its length to grow lines of full-grown trees.

**Hewn timber** A log that has had its sides shaped using an axe.

**Honed** A sharp edge that is made razor-sharp using sharpening stones.

**Hook pins** See teepins.

**Horn** See ear.

**Horse** See trestle.

**Housing** An area of a timber that is recessed or cut away to receive another piece of wood.

**Hypotenuse** The longest side of a right-angle triangle.

**Impact driver** A hammer-action cordless screwdriver.

**Ink-line** A long fine string daubed in ink to be stretched and snapped onto a timber to leave a straight line of ink on its surface.

**Joint** The intersection between two or more pieces of wood that holds them all together.

**Joist** Horizontal timbers that support floorboards.

**Jowl post** A vertical post that flares to a greater width at its head.

**Jowl post chin** The small, shaped step of timber between the sloping angle at the base of the jowl and the main surface of the post.

**Kelly kettle** A kettle with an in-built small firebox to burn twigs for heating water.

**Kerf** The saw-width cut made when sawing timber.

**King post** A post in a roof truss rising from a tie beam and into which the heads of the principal rafters are jointed.

**Knots (of a tree)** The remnants of the base of a branch embedded within a tree stem, formed as the tree stem grows outwards and envelopes the branch.

**Lap joint** When one piece of wood laps on top of another.

**Late wood** The part of each annual ring growing from late spring and summer that has smaller or fewer vessels for transporting sap than early wood and also includes fibres that give strength to the tree stem.

**Lathe** A framework for holding a spinning piece of wood which is then cut with a chisel held stationary. Either powered by a motor or a foot-powered treadleboard.

**Lay-up** Two or more timbers from a frame arranged together, usually on trestles, to mark and cut joints.

**Lignin** The organic chemical compound giving rigidity to the vessels within timber.

**Lime mortar** A mixture of sand and calcium hydroxide.

**Log list** A list of numbered logs to be milled giving the diameter, length, shape and condition.

**Maul** A heavy wooden cudgel like a stunted baseball bat.

**Medullary rays** Narrow sheets of tissue in the stem of trees radiating from the centre.

**Mid-strey** The bay of a barn with the large doors, usually central.

**Mid-rail** A horizontal beam in a wall frame, usually halfway up.

**Milling** Sawing logs or timber along their lengths to produce flat-sided timber.

**Milling list** A list of the sizes, lengths and quantities of timbers from the cutting list matched to identified logs from the log list.

**Mortice** The housing cut out of a timber to receive the shaped end of another piece of wood.

**Mortice cheek** The back and top sides of a mortice box roughly parallel with the face of the timber.

**Mortice gauge** A tool to mark one or two scratched lines on a timber surface parallel to an edge.

**Mortice tail** See tail.

**Nave** The central space of a building.

**Nib** A small projecting piece of wood or other material.

**Nose** The parts of a mortice and a tenon that press upon each other inside the joint in a sideways direction.

**Offset hole** The peg hole in a tenon that is deliberately misaligned to the peg hole in the mortice timber.

**Offset line** An ink or pencil line made parallel to a timber edge by a predetermined amount at each end.

**Out-shut** Part of a structure projecting from a wall whereby the sloping roof also extends outwards.

**Packer** A thin strip of wood or plywood.

**Parbuckling** A system of rolling tree logs sideways from the ground up onto the bed of a truck or trailer.

**Paring** Removing thin slivers of wood from a surface.

**Peg** A small length of timber knocked into a peg hole to hold a joint together.

**Phloem** See bast.

**Pikestaff** A long, strong pole for pushing the frame upright during the raising.

**Pit saw** A long saw used vertically to cut along the line of a log with a 'T' handle at the top and a dismountable handle at the bottom.

**Pith** The spongy strand in the centre of a tree stem following the trace of the growth of the leading tip of the tree or branch.

**Plane** A continuous flat two-dimensional surface.

**Plinth wall** The base wall supporting a timber frame.

**Plumb-bob** A weight on the end of a string.

**Plumbed** When a face or side of a timber is adjusted to become vertical.

**Podger** See teepin.

**Points** See dividers.

**Pony** See trestle.

**Post** A vertical timber in a frame that supports a horizontal timber above it.

**Primary frame** The timbers of a frame that carry the structural loads that keep the frame upright as opposed to those timbers carrying the roof covering, wall cladding and floorboards.

**Primary timbers** Lengths of wood forming the primary frame.

**Principal rafter** A rafter that is part of the primary frame and usually integral to a cross-frame.

**Pulley block** One or more wheels arranged around a common spindle within a supporting case for ropes to run around.

**Purlin** A roof timber running horizontally between cross-frames that supports common rafters.

**Purlin scarf** A longitudinal joint connecting two purlin lengths.

**Quarter sawn** An alignment of the milling blade to pass through the middle of a log. This results in the annual rings being at right angles to the surface of the sawn timber and visually exposes the medullary rays.

**Queen strut** A primary timber in the roof structure of a cross-frame rising from a tie beam to support the principal rafter.

**Radial shake** A crack starting on the perimeter of a log that extends inwards as the timber dries.

**Rafter foot** The bottom end of a rafter.

**Rafters** The sloping timbers of a roof running from a wall to a ridge.

**Rail** A horizontal timber within a wall frame.

**Ratchet straps** Webbing straps, sometimes called lorry straps, with a small ratchet winch mechanism to pull the webbing tight.

**Rhizome** An underground stem of a fungus.

**Ridge** The topmost timber of a roof, running longitudinally and supporting the rafters.

**Ridge purlin** A ridge timber that rafters rest onto, rather than a ridge board that the rafters abut.

**Rift sawn** An alignment of the milling blade to pass through a log at approximately a quarter or three quarters of its diameter. This results in the annual rings being at 45 degrees to the surface of the sawn timber.

**Ripsaw** A handsaw with its teeth designed to cut along the length of a timber.

**Rising brace** A primary timber that diagonally rises from a horizontal timber to a vertical post and prevents the frame from rocking to and fro.

**Roll mark** A mark showing the position on a face or side of a timber that becomes vertical or horizontal when the timber is rolled around its axis.

**Sap** Water with dissolved minerals and nutrients that is drawn upwards from the roots to the leaves through the vessels of sapwood.

**Sapwood** The layers of vessels and fibres in the outer annual rings of a tree immediately under the cambium, also called xylem.

**Sawmill** A machine for cutting logs and timber lengthwise.

**Scantling** Temporary or lightweight timbers providing propping or vertical support.

**Scarf** A timber joint connecting two timbers end to end.

**Scotch** A housing cut out of the top surface of the horizontal timber at the top of a wall to support the foot of a rafter.

**Scratch awl** A small tool with a metal spike that is sharpened to a point to scratch accurate marks on the surface of timber.

**Scribe** To transfer positions of faces and edges from one timber to another.

**Second-rate timber** Wood that has physical flaws like large knots that render it weaker in a range of uses.

**Shave-horse** A wooden bench-like structure to sit astride with a vice-like bar that grips a piece of wood so that it can be worked on. Used to make pegs with a drawknife.

**Shaw** See hedge row.

**Sheath** An extra layer of material, such as plywood, fixed to timbers to strengthen them.

**Sheer legs** A braced pair of sloping lengths of timber fixed at the top to support a pulley block and rope. Used in place of a crane.

**Shoulders** The surface of wood on one or two sides of a tenon in a mortice and tenon joint that take the weight and compression forces from the timber it is connected to.

**Side axe** An axe with a bevel on only one side, so that it is like a chisel in cross-section.

**Sidings** The thin boards produced when a log is squared then reduced to the desired timber section size.

**Slab** The first bark-covered length of wood produced when the sides of a log are milled.

**Slab rail** Long, even and regularly shaped slabs produced through milling.

**Sling brace** A long arch brace that rises from a post to a collar, with the tie beam cut short and jointed into it.

**Slop** The small gap between a tenon and the sides of the mortice box.

**Snapping** A line. See ink-line.

**Softwood** Species of trees that do not lose their leaves in winter. Also called conifers.

**Sole plate** The horizontal timber at the base of a wall frame that sits on a plinth wall. Sometimes called a sill.

**Spars** The local name given rafters and the small sapling trees that they are made from.

**Spike** A nail or sharp point of metal.

**Spinney** A small isolated group of trees.

**Spiral grain** Alignment of the vessels and fibres of the woody material, running like a corkscrew along the length of a tree.

**Spire** The part of a tree that continues straight upwards past the first large branch.

**Spirit level** A tool with an air bubble in a tube set in a length of wood or metal to signify vertical or horizontal orientation of the surface it is held against.

**Splay** A longitudinally sloping surface.

**Spring** A curve in a timber that moves in and out of the face of a frame.

**Sprocket** A piece of timber attached to the bottom of a rafter and extending outwards over the top of the wall.

**Squaring up** Shuffling two or more timbers until they are at right angles with each other.

**Star shake** Long, continuous splits radiating from the pith of a tree as the tree grows.

**Stem** A continuous length of timber in the round, such as a tree trunk, coppice regrowth or branch of a tree.

**Stickering sticks** Small section battens laid in between boards to allow air ventilation.

**Stool** See trestle, or, the living base of a coppiced tree that sends out new shoots when the previous growth has been harvested back to within a foot or so from the ground.

**String line** A long string pulled tight to provide a reference line.

**Strop** A webbing strap with hoops at its ends.

**Struts** Timbers to take compression loads.

**Stub-tenons** Short tenons that don't use pegs to hold them in place.

**Stub-tie** A shortened tie beam that spans from the topmost timber of a wall but doesn't span as far as the opposite wall.

**Studs** Vertical timbers fitted to the structural frame to support a cladding material, boarding or internal finishes.

**Subsoil** The layer of ground beneath topsoil.

**Swing-shovel excavator** A large machine for digging.

**Tails** The lines used to mark joints, such as mortice and tenons that over-shoot to extend beyond the end of the joint itself.

**Tannins** Natural preservatives found in the heartwood of some trees.

**Tapered peg** A peg that reduces in diameter from one end to the other.

**Teazle** The part of a jowl post that extends up and round the side of timber that the rest of the post is joined to.

**Teazle tenon** The tenon at the end of the teazle.

**Teepin** A metal dowel with a tapered end and a 'T' handle used as temporary pegs in mortice and tenon joints in a frame assembly.

**Tenon** The shaped end of a piece of wood that fits into a mortice.

**Tenon's tongue** A protruding length of shaped timber in a tenon

**Thinnings** Trees and saplings that are removed from a wood to allow the remaining trees to continue to grow more fully.

**Throat** The part of a mortice or a tenon that does not bear on another surface in a sideways direction within the joint.

**Throat line** A dashed line marked in the mortice timber face to show position of the throat.

**Through-sawn** A way of milling a log where all the cuts are parallel with each.

**Tie beam** The large timber of a cross-frame spanning from the top of the front wall to the top of the rear wall.

**Timber trolley** A two-wheeled trolley used to transport timbers around the framing yard, usually with its bed at about waist height.

**Tirfor winch** A manufacturer's brand of wire winch.

**Top chord** See chord.

**Top plate** The topmost horizontal timber in a wall frame.

**Top-of-top-plate line** A line marked on a tie beam between the positions of the tops of the top plates at each of its ends.

**Topping-out** Celebrating the last timber that goes up during the raising of a timber frame.

**Topsoil** The topmost layer of soil comprising mineral particles, organic matter, water, air, microorganisms and invertebrates.

**Torrak** Wood from dead standing trees, though not necessarily rotten or unsound.

**Trestle** A horizontal timber with three or four legs at a convenient height to sit on or support a timber.

**Truss** The part of a cross-frame supporting the roof.

**Try square** A flat, straight-edged piece of metal fixed at a right angle to another.

**Tumbling** A system of marking the positions and bevels for shoulders of studs without measurements.

**Tushing** Dragging logs along with one end sliding on the ground.

**Tyloses** Minute knobs within the vessels of timber produced by the tree to block them and render them non-porous and more resistant to fungal decay.

**Under-cut** Cutting a timber at an angle that slopes underneath the marked line.

**Wane** The undulating edge of a tree's outer surface under the bark and bast.

**Waney-edged** The natural shape at the edge of a piece of wood that has been milled with the edge of the tree intact.

**Wassailing** A tradition of singing and making loud noises with percussion instruments to wake up apple trees in winter.

**Waste side** The side of a marked line on a timber that will not be part of the finished piece.

**Weatherboard** Overlapping external horizontal cladding boards.

**Wedge** A tapered wooden block.

**Wind** Pronounced 'wined'. The twist in a piece of timber from one end to the other, visible along its surface.

**Wind brace** The diagonal braces in a roof structure rising from a principal rafter to a purlin.

**Working drawings** Scaled drawings with all relevant dimensions and notes. Used on-site as the reference drawing.

**Wrack** Unsound or rotten timber.

**Xylem** The vessels and fibres that form the woody material of trees and transport water and minerals as well as holding them upright. Also known as sapwood and heartwood.

**Yoke** The small horizontal timber at the top of a pair of rafters that supports a longitudinal ridge timber.

# Resources

## Full Framing Details

For a comprehensive guide to each of the lay-ups and step-by-step instructions for scribing and cutting the different joints for a classic Hertfordshire elm barn, an illustrated guide is available via the author. Visit www.elmbarncarpentryhandbook.co.uk for more details.

## Recommended Reading

The following are recommended as informative and enjoyable reading:

John Evelyn *Sylva: A Discourse on Forest Trees and the Propagation of Timber in His Majesty's Dominion.* J. Martyne, J. Mackock and J. Allestry for the Royal Society, London, 1664.

Batty Langley *A Sure and Easy Method of Improving Estates.* F. Noble, London, 1740.

William Ellis *The Timber Tree Improved: or, The Best Practical Methods of Improving Different Lands with Proper Timber.* 2nd edition J. and J. Fox, T. Cooper and E. Withers, London, 1742.

William Boutcher *A Treatise on Forest-trees.* William Wilson and John Exshaw, Dublin, 1776.

Thomas Laslett *Timber and Timber Trees.* Macmillan and Co., London, 1875.

J. Davies *Building Timbers and Architect's Specifications.* Alfred Haworth & Co., Ltd., London, 1910.

Henry John Elwes and Augustine Henry *The Trees of Great Britain and Ireland, Volume 7.* R. and R. Clark Ltd, Edinburgh, 1913.

George Sturt *The Wheelwright's Shop.* Cambridge University Press, 1923.
Walter Rose *The Village Carpenter.* Cambridge University Press, 1937.
George Orwell *Animal Farm.* Martin Secker and Warburg Ltd, 1945.
Herbert L. Edlin *Woodland Crafts in Britain.* David and Charles, Newton Abbot, 1949.
Herbert L. Edlin *What Wood is That?* Thames and Hudson Ltd., London, 1969.
Richard Harris *Discovering Timber-Framed Buildings.* Shire Publications Ltd, Aylesbury, 1978.
Cecil A. Hewett *English Historic Carpentry.* Phillimore and Co. Ltd, 1980.
R.H. Richens *Elm.* Cambridge University Press, 1983.
Mike Abbott *Green Woodwork.* Guild of Master Craftsman Publications Ltd, Lewes, 1989.
Jack A. Sobon *Build a Classic Timber-Framed House.* Storey Books, Vermont, 1994.
Michael Drury *Wandering Architects.* Shaun Tyas, Lincolnshire, 2000.
Jim Coutts Smith *George Orwell in Wallington.* Mardleybury Publishing, 2010.
Tom Williamson, Gerry Barnes and Toby Pillatt *Trees in England.* University of Hertfordshire Press, 2017.
Peter Sell and Gina Murrell *Flora of Great Britain and Ireland, Volume 1.* Cambridge University Press, 2018.
Joshua A. Klein *Another Work is Possible.* Mortise and Tenon Inc., Sedgewick, 2020.

## Relevant Organisations

The Carpenters' Fellowship (UK) info@carpentersfellowship.co.uk
The Timber Framers Guild (USA) info@tfguild.org
Charpentiers Sans Frontières contact@charpentiers-sans-frontieres.com
Weald and Downland Living Museum office@wealddown.co.uk
Chiltern Open Air Museum enquiries@coam.org.uk
Guedelon Castle (France) guedelon@guedelon.fr
The Conservation Foundation (UK) info@conservationfoundation.co.uk

## Tool Suppliers

Apart from your own local market stalls and network of tool suppliers, the following can provide all the tools needed for timber framing by hand:

Classic Hand Tools: sales@classichandtools.com, 01473 784 983
(new and second-hand tools)
Timber Framing Tools: info@tftools.co.uk, 01225 929 1740
(new framing tools)
Bristol Design: tools@bristol-design.co.uk, 0117 929 1740
(second-hand tools)
G and M Tools: sales@gandmtools.co.uk, 01903 892 510
(second-hand tools)
Greenwood Direct Ltd: tim@greenwood-direct.co.uk, 01249 782 100
(good quality axes)
Tewkesbury Saw Ltd: sales@tewkesburysaw.co.uk, 01684 293 092
(saw doctor)
Totnes Classic Handtools: tomwid909@gmail.com, 07825 136699
(market stall selling quality used hand tools)

# About the Author

Jonathan Sampson

**Robert J. Somerville** grew up in rural Kent during the 1960s. A childhood spent in the woods and hills of the North Downs and then the orchards and marshes of the Little Stour valley inspired his deep love for the natural world. After studying engineering and architecture at Cambridge University, Somerville went on to run a design and building business in Devon, utilising local wood, stone and earth.

Upon moving to Hertfordshire with his wife, Lydia, he began working with local woodland owners and foresters to source local elm timber and then build and raise timber frames by hand, with the help of volunteers known as the Barn Club – a group formed to teach, practise and celebrate skilled rural craftsmanship.

Somerville lives with Lydia and their daughter in a self-built eco-house in an idyllic smallholding.